The Human Brain during the Second Trimester 160- to 170-mm Crown-Rump Lengths

This ninth of 15 short atlases reimagines the classic 5-volume *Atlas of Human Central Nervous System Development*. This volume presents serial sections from specimens between 160 mm and 170 mm with detailed annotations. An introduction summarizes human CNS developmental highlights around 5 months of gestation. The Glossary (available separately) gives definitions for all the terms used in this volume and all the others in the *Atlas*.

Key Features

- Classic anatomical atlas
- Detailed labeling of structures in the developing brain offers updated terminology and the identification of unique developmental features, such as germinal matrices of specific neuronal populations and migratory streams of young neurons
- Appeals to neuroanatomists, developmental biologists, and clinical practitioners
- A valuable reference work on brain development that will be relevant for decades

ATLAS OF
HUMAN CENTRAL NERVOUS SYSTEM DEVELOPMENT
Series

The Human Brain during the Second Trimester 160- to 170-mm Crown-Rump Lengths

Atlas of Human Central Nervous System Development, Volume 9

Shirley A. Bayer
Joseph Altman

CRC Press
Taylor & Francis Group
Boca Raton London New York

CRC Press is an imprint of the
Taylor & Francis Group, an **informa** business

First edition published 2024
by CRC Press
6000 Broken Sound Parkway NW, Suite 300, Boca Raton, FL 33487-2742

and by CRC Press
4 Park Square, Milton Park, Abingdon, Oxon, OX14 4RN

CRC Press is an imprint of Taylor & Francis Group, LLC

LCCN no. 2022008216

ISBN: 978-1-032-21943-1 (hbk)
ISBN: 978-1-032-21942-4 (pbk)
ISBN: 978-1-003-27068-3 (ebk)

DOI: 10.1201/9781003270683

Typeset in Times Roman
by KnowledgeWorks Global Ltd.

Access the Support Material: www.routledge.com/9781032219431

CONTENTS

ACKNOWLEDGMENTS

We thank the late Dr. William DeMyer, pediatric neurologist at Indiana University Medical Center, for access to his personal library on human CNS development. We also thank the staff of the National Museum of Health and Medicine that were at the Armed Forces Institute of Pathology, Walter Reed Hospital, Washington, D.C. when we collected data in 1995 and 1996: Dr. Adrianne Noe, Director; Archibald J. Fobbs, Curator of the Yakovlev Collection; Elizabeth C. Lockett; and William Discher. We are most grateful to the late Dr. James M. Petras at the Walter Reed Institute of Research who made his darkroom facilities available so that we could develop all the photomicrographs on location rather than in our laboratory in Indiana. Finally, we thank Chuck Crumly, Neha Bhatt, Kara Roberts, Michele Dimont, and Rebecca Condit for expert help during production of the manuscript.

AUTHORS

Shirley A. Bayer received her PhD from Purdue University in 1974 and spent most of her scientific career working with Joseph Altman. She was a professor of biology at Indiana-Purdue University in Indianapolis for several years, where she taught courses in human anatomy and developmental neurobiology while continuing to do research in brain development. Her lengthy publication record of dozens of peer-reviewed, scientific journal articles extends back to the mid 1970s. She has co-authored several books and many articles with her late spouse, Joseph Altman. It was her research (published in *Science* in 1982) that proved that new neurons are added to granule cells in the dentate gyrus during adult life, a unique neuronal population that grows. That paper stimulated interest in the dormant field of adult neurogenesis.

Joseph Altman, now deceased, was born in Hungary and migrated with his family via Germany and Australia to the US. In New York, he became a graduate student in psychology in the laboratory of Hans-Lukas Teuber, earning a PhD in 1959 from New York University. He was a postdoctoral fellow at Columbia University, and later joined the faculty at the Massachusetts Institute of Technology. In 1968, he accepted a position as a professor of biology at Purdue University. During his career, he collaborated closely with Shirley A. Bayer. From the early 1960s-2016, he published many articles in peer-reviewed journals, books, monographs, and free online books that emphasized developmental processes in brain anatomy and function. His most important discovery was adult neurogenesis, the creation of new neurons in the adult brain. This discovery was made in the early 1960s while he was based at MIT, but was largely ignored in favor of the prevailing dogma that neurogenesis is limited to prenatal development. After Dr. Bayer's paper proved new neurons are added to granule cells in the hippocampus, Dr. Altman's monumental discovery became more accepted. During the 1990s, new researchers "rediscovered" and confirmed his original finding. Adult neurogenesis has recently been proven to occur in the dentate gyrus, olfactory bulb, and striatum through the measurement of Carbon-14—the levels of which changed during nuclear bomb testing throughout the 20th century—in postmortem human brains. Today, many laboratories around the world are continuing to study the importance of adult neurogenesis in brain function. In 2011, Dr. Altman was awarded the Prince of Asturias Award, an annual prize given in Spain by the Prince of Asturias Foundation to individuals, entities, or organizations globally who make notable achievements in the sciences, humanities, and public affairs. In 2012, he received the International Prize for Biology - an annual award from the Japan Society for the Promotion of Science (JSPS) for "outstanding contribution to the advancement of research in fundamental biology." This Prize is one of the most prestigious honors a scientist can receive. When Dr. Altman died in 2016, Dr. Bayer continued the work they started over 50 years ago. In her late husband's honor, she created the Altman Prize, awarded each year by JSPS to an outstanding young researcher in developmental neuroscience.

INTRODUCTION

A. Specimens and Organization

Volume 9 in the *Atlas of Human Central Nervous System Development* series illustrates development of the 5-month-old human brain in the middle second trimester (originally presented in Bayer and Altman 2005). Throughout the diencephalon, midbrain, pons, and medulla differentiation is rapidly progressing and appears mature at low magnification. The telencephalon (cerebral cortex, basal ganglia) continue generation of microneurons, and complex migration patterns appear, especially in the cerebral cortex. Immature features continue to predominate in the cerebellum where neurons are still migrating and its secondary germinal matrix, the external germinal layer, continues generation of basket, stellate, and granule cells.

Grayscale photographs of Nissl-stained sections of three normal brains in the Yakovlev Collection[1] are illustrated. **Part II** presents Y27-60 (160-mm crown-rump length) cut in the sagittal plane. Sagittal plates are ordered from medial to lateral; the anterior part of each section is facing to the left, posterior to the right. **Part III** presents Y17-63 (165-mm crown-rump length) cut in the frontal plane. Frontal plates are presented in serial order from rostral to caudal; the dorsal part of each section is toward the top of the page, the ventral part at the bottom, and the midline is in the vertical center of each section. **Part IV** presents Y391-62 (170-mm crown-rump length) cut in the horizontal plane. Horizontal plates are presented in serial order from top (dorsal/superior) to bottom (ventral/inferior; the anterior part of each section is on the left, the posterior part on the right, and the midline is in the horizontal center of each section. **Part A** of each plate on the left page shows the full-contrast photograph without labels; **Part B** shows low-contrast copies of

the same photograph on the right page with superimposed outlines of structures and unabbreviated labels. The *low-magnification plates* show entire sections to identify large structures of the brain. The *high-magnification plates* feature enlarged views of the brain core to identify smaller structures. A few *very-high-magnification plates* show the cerebral cortex in great detail. As in other volumes, transient structures are labeled in ***italics***. During fixation, shrinkage introduced artifactual infolding of the cerebral cortex in some specimens. During dissection, embedding, cutting, and staining, some of the sections illustrated were torn. Both artifacts and processing damage are usually outlined with *dashed lines* in part **B** of each plate.

B. Developmental Highlights

The focus is on the ***stratified transitional field*** (***STF***) because it reaches its peak differentiation in the 5th month. A parasagittal slice of Y27-60 (**Fig. 1**) shows changes in the ***STF*** between sensory and motor areas. A detailed diagram explains hypothetical happenings in ***STF3***. Illustrations are modified from our book, *Development of the Human Neocortex* (Altman and Bayer, 2015), which is available free online at neurondevelopment.org.

1. The *Yakovlev Collection* (designated by a **Y** prefix in the specimen number) is the work of Dr. Paul Ivan Yakovlev (1894–1983), a neurologist affiliated with Harvard University and the AFIP. Throughout his career, Yakovlev collected many diseased and normal human brains. He invented a giant microtome that was capable of sectioning entire human brains. Later, he became interested in the developing brain and collected many human brains during the second and third trimesters. The normal brains in the developmental group were cataloged by Haleem (1990) and were examined by us during 1996 and 1997. The collection was moved to the National Museum of Health and Medicine when the Armed Forces Institute of Pathology (AFIP) closed and is still available for research.

REFERENCES

Altman J, Bayer SA. (2015) *Development of the Human Neocortex*. Ocala, FL, Laboratory of Developmental Neurobiology, neurondevelopment.org.

Bayer SA, Altman J (2005) *Atlas of Human Central Nervous System Development*, Volume 3: *The Human Brain during the Second Trimester*. Boca Raton, FL, CRC Press.

Haleem M (1990) *Diagnostic Categories of the Yakovlev Collection of Normal and Pathological Anatomy and Development of the Brain*. Washington, D.C. Armed Forces Institute of Pathology.

Retzius, G. (1896) *Das Menschenhirn: Studien in der makroskopischen Morphologie*. Vols. 1 and 2. Stockholm: Königliche Buchdruckerei.

LAYERS IN THE STRATIFIED TRANSITIONAL FIELD (STF)

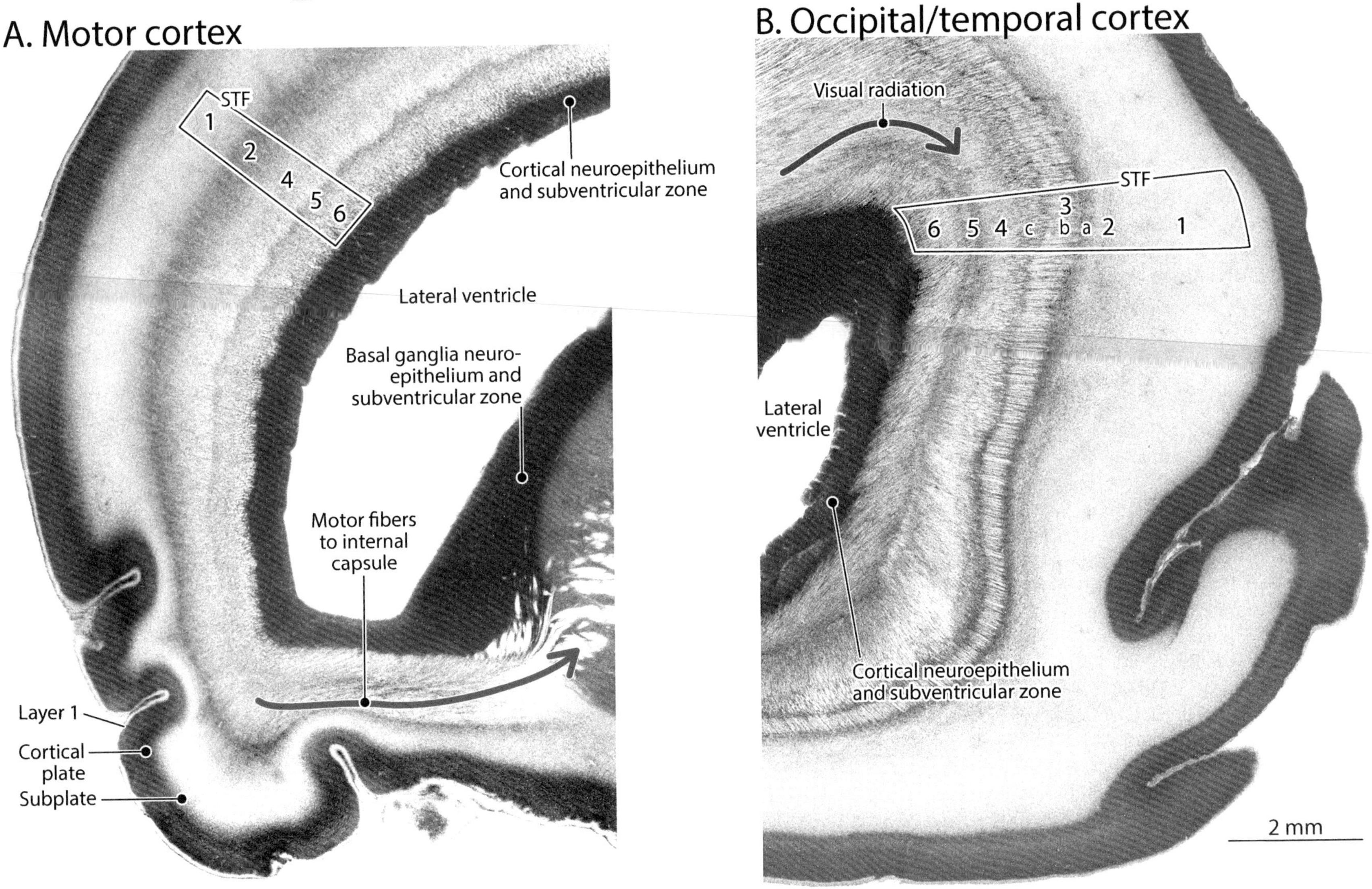

A. Motor cortex

B. Occipital/temporal cortex

STF
1
2
4
5 6

Cortical neuroepithelium
and subventricular zone

Lateral ventricle

Basal ganglia neuro-
epithelium and
subventricular zone

Motor fibers
to internal
capsule

Layer 1
Cortical
plate
Subplate

Visual radiation

STF
6 5 4 c b a 2 1
3

Lateral
ventricle

Cortical neuroepithelium
and subventricular zone

2 mm

Figure 1. Parasagittal slices through the motor cortex (**A**, Section 210) and the occipital/temporal cortex (**B**, Section 101) of Y27-60, the specimen in **Part II**, show the sensory motor differences in *STF* layering. *STF4* is the main output conduit of the motor cortex (*red arrow* in **A**) and input conduit for cortical afferents, most likely visual fibers from the thalamic lateral geniculate body (*purple arrow* in **B**). Both areas have future callosal fibers in *STF6* that are heavily infiltrated by cells migrating from the cortical germinal zones that are still active due to the prolonged cortical development in humans. *STF2* is denser in motor cortex because it is a likely sojourn zone for the multitude of cells in Layer V; *STF2* is much thinner in sensory cortex because Layer V is less prominent. *STF5* is thinner in motor (fewer cells in Layer IV) vs sensory cortex (many cells in Layer IV) because this is where neurons of Layers IV-II sojourn before migrating to the cortical plate. Neurons destined for Layers IV-II in sensory areas migrate through a three-part honeycomb matrix in *STF3*, absent in motor areas, where we hypothesize (**Figure 2**, facing page) cell specification and initial anatomical connections are made. (Modified Figure 21 in Altman and Bayer, 2015.)

HYPOTHETICAL FUNCTIONAL ORGANIZATION OF THE STRATIFIED TRANSITIONAL FIELD (STF) IN THE VISUAL CORTEX

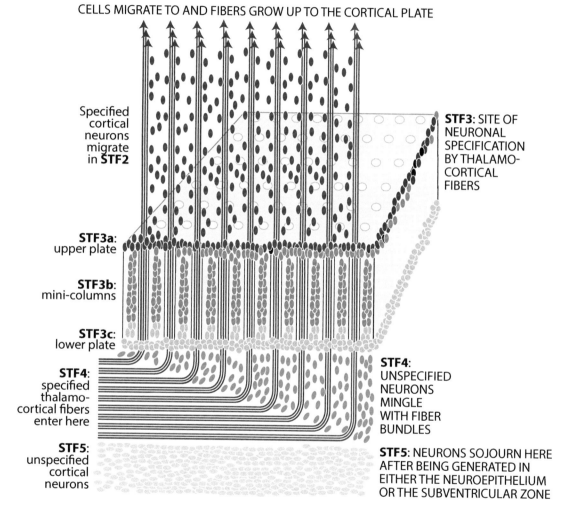

Figure. 2 A diagram of the organization and hypothetical function of *STF3* in the developing visual cortex. The two horizontal cell-dense bands (*STF3a* and *STF3c*) form an upper plate and a lower plate parallel to the cortical surface, with vertical minicolumns of cells and fibers between them (*STF3b*). We postulate that the segregated fiber bundles from the lateral geniculate nucleus (in *STF4*) carry specific information, such as location of retinal origin, on-center or off-center excitability, etc., and that these fibers interact with the nonspecified neurons moving out of *STF5*. As neurons intermingle with the thalamic fibers in *STF4*, they then pass through *STF3c*, *STF3b*, and *STF3a*. We postulate interactions between the cells and fibers during the transit. The cells then enter *STF2* as specified neurons, possibly migrating to specific regions in the cortical plate with their associated fibers. (Figure 23 in Altman and Bayer, 2015.)

PART II: Y27-60
CR 160 mm (GW 20)
Sagittal

This specimen is a premature female neonate that died one hour after birth from the aspiration of amniotic fluid (Yakovlev case number RPSL W-27-60; here referred to as Y27-60) with a crown-rump length (CR) of 160-mm estimated to be at gestational week (GW) 20. The brain was cut in the sagittal plane in 724 sections (35-μm thick) and is classified as a Normative Control in the Yakovlev Collection (Haleem, 1990). Since there is no photograph of this brain before it was embedded and cut, we turned to the comprehensive atlas that Retzius published in 1896 showing whole fetal brains in medial, lateral, superior, and inferior views and midline sagittally cut brains. **Figure 3** (taken from Retzius (1896), shows the midline sagittal surface of a brain from a specimen that is at the same age as Y27-60. Photographs of 6 Nissl-stained sections from the right hemisphere are shown at low-magnification in **Plates 1-6**. High-magnification views of different regions of the cerebral cortex are shown in **Plates 7** and **8**, and of the midline cerebellar cortex in **Plates 9** and **10**. A midline section is not shown because there is much damage in diencephalic and mesencephalic regions.

In the telencephalon, the ***cortical neuroepithelium/subventricular zone*** is bordered by sharply defined layers of the ***stratified transitional fields (STF)*** in all lobes of the cerebral cortex. A single lateral sagittal section shows the differences in the ***STF*** between future sensory and future motor areas (**Fig. 1**). High-magnification views show sharp delineations of the various ***STF*** layers and the cortical plate. The ***lateral migratory stream*** in the cerebral cortex has streams of cells percolating through the claustrum, endopiriform nucleus, external capsule, and uncinate fasciculus that invade the insular cortex, primary olfactory cortex, temporal cortex, and basolateral parts of the amygdaloid complex. The ***ammonic migratory stream*** is more definite in the hippocampus. Stem cells in the ***subgranular zone*** generate dentate granule cells. There is a very thick ***neuroepithelium/subventricular zone*** in the nucleus accumbens and striatum where neurons (and glia) are generated. In this sagittally sectioned brain, the striatal neuroepithelium and subventricular zone can be more easily subdivided into anterolateral, anteromedial, and posterior parts than in frontally sectioned brains. A ***glioepithelium/ependyma*** lines the ventricle in the septum. Generally, brainstem structures appear more mature than in the telencephalon, but lateral sections of the thalamus show a darkly staining stream of migrating cells heading toward the lateral geniculate body.

The cerebellum in this specimen shows well-defined lobules in the vermis and hemisphere. The dentate nucleus is not yet laminated. The entire surface of the cerebellar cortex is covered by the prominent ***external germinal layer (egl)*** that is actively producing basket, stellate, and granule cells. Lamination in the cortex is immature, except for a thin molecular layer beneath the ***egl***. Most cortical neurons, including Purkinje cells, are still migrating rather than settling. However as shown in the high-magnification views of the cortex, there are more settled Purkinje cells in the anterior vermis (culmen) than in the central vermis (declive). The ***germinal trigone*** is large at the base of the nodulus and along the floccular peduncle; choroid plexus cells and glia are originating here.

GW20 MIDLINE SAGITTAL VIEW

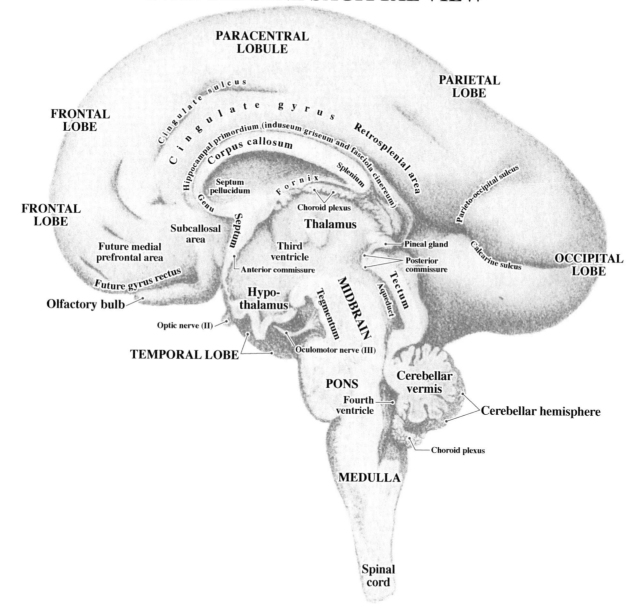

Figure 3. Midline sagittal view of a GW20 brain with major structures in the cerebral hemispheres and brainstem labeled. (This is Figure 19 in Table 4, Volume 2, Retzius, 1896.)

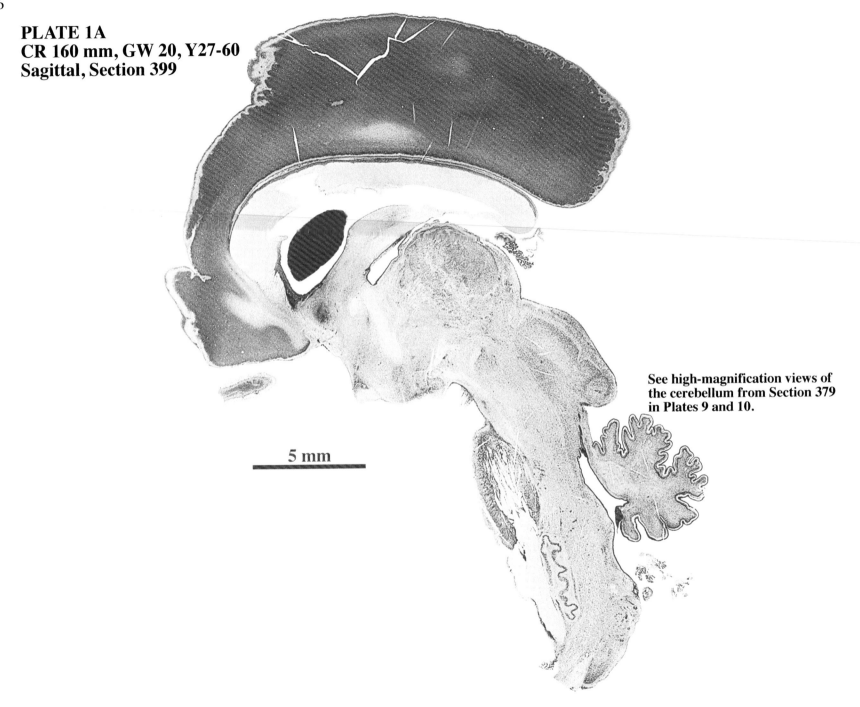

6

PLATE 1A
CR 160 mm, GW 20, Y27-60
Sagittal, Section 399

See high-magnification views of
the cerebellum from Section 379
in Plates 9 and 10.

5 mm

7

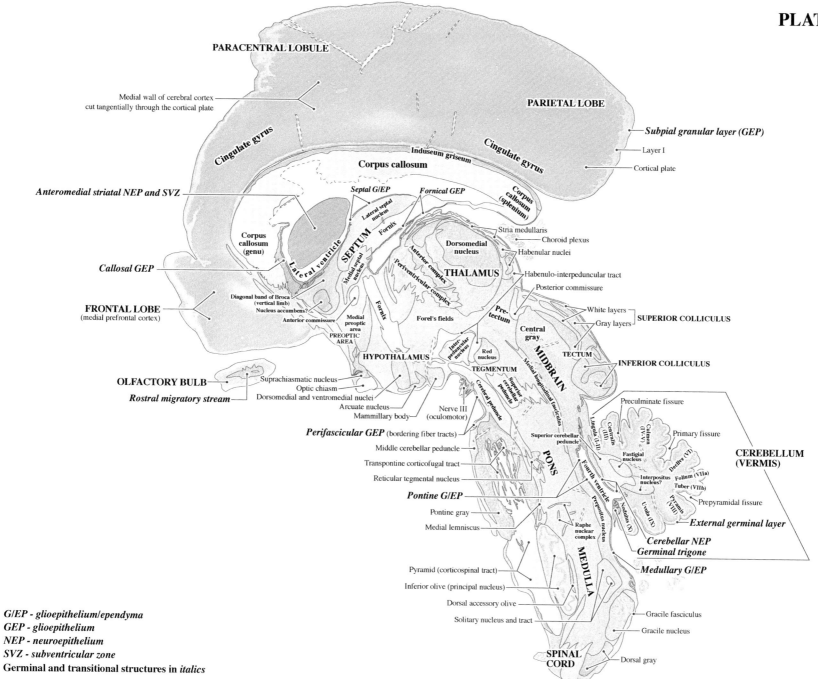

PARACENTRAL LOBULE

Medial wall of cerebral cortex
cut tangentially through the cortical plate

PARIETAL LOBE

Cingulate gyrus

Cingulate gyrus

Subpial granular layer (GEP)

Induseum griseum

Layer I

Corpus callosum

Cortical plate

Septal G/EP

Fornical GEP

Corpus callosum (splenium)

Anteromedial striatal NEP and SVZ

Lateral septal nucleus

Fornix

Stria medullaris

Choroid plexus

Dorsomedial nucleus

Habenular nuclei

Corpus callosum (genu)

SEPTUM

Anterior complex

THALAMUS

Habenulo-interpeduncular tract

Callosal GEP

Medial septal nucleus

Periventricular complex

Posterior commissure

Lateral ventricle

Diagonal band of Broca (vertical limb)

White layers

Pre-tectum

FRONTAL LOBE
(medial prefrontal cortex)

Nucleus accumbens?

Fornix

Forel's fields

Gray layers

SUPERIOR COLLICULUS

Anterior commissure

Medial preoptic area

Central gray

TECTUM

PREOPTIC AREA

HYPOTHALAMUS

Inter-peduncular nucleus

MIDBRAIN

INFERIOR COLLICULUS

Red nucleus

TEGMENTUM

Medial longitudinal fasciculus

OLFACTORY BULB

Suprachiasmatic nucleus

Optic chiasm

Cerebral peduncle

Superior cerebellar peduncle

Rostral migratory stream

Dorsomedial and ventromedial nuclei

Nerve III (oculomotor)

Arcuate nucleus

Mammillary body

Preculminate fissure

Perifascicular GEP (bordering fiber tracts)

Superior cerebellar peduncle

Lingula (I-II)

Centralis (IV-V)

Primary fissure

Middle cerebellar peduncle

Culmen (IV-V)

Transpontine corticofugal tract

Fastigial nucleus

Declive (VI)

CEREBELLUM (VERMIS)

Reticular tegmental nucleus

PONS

Fourth ventricle

Interpositus nucleus?

Folium (VIIa)

Tuber (VIIb)

Pontine G/EP

Prepositus nucleus

Prepyramidal fissure

Pontine gray

Nodulus (X)

Pyramis (VIII)

External germinal layer

Medial lemniscus

Raphe nuclear complex

Uvula (IX)

Cerebellar NEP
Germinal trigone

MEDULLA

Medullary G/EP

Pyramid (corticospinal tract)

Inferior olive (principal nucleus)

Dorsal accessory olive

Gracile fasciculus

Solitary nucleus and tract

Gracile nucleus

SPINAL CORD

Dorsal gray

G/EP - glioepithelium/ependyma
GEP - glioepithelium
NEP - neuroepithelium
SVZ - subventricular zone
Germinal and transitional structures in *italics*

8

PLATE 2A
CR 160 mm
GW 20
Y27-60
Sagittal
Section
439

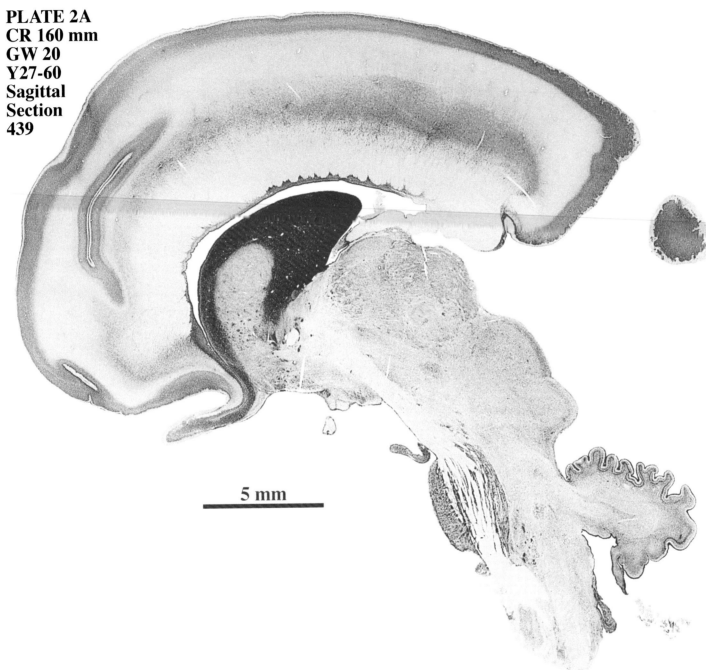

5 mm

LAYERS OF THE CORTICAL
STRATIFIED TRANSITIONAL
FIELD (STF)

STF1—Superficial fibrous
layer with an early
developmental stage *(t1)*
when many cells are
migrating through it, followed
by a late stage *(t2)* with sparse
cells. Endures as the
subcortical white matter.

STF4—Complex middle
layer with three
developmental stages:
t1– fibrous layer without
interspersed cells;
t2– cells and fibers
intermingle to form striations;
t3– fibers endure in the deep
white matter.

STF5—Deep cellular layer,
the first sojourn zone to
appear outside the germinal
matrix.

STF6—Late-forming deep
layer of callosal fibers outside
the germinal matrix.

9

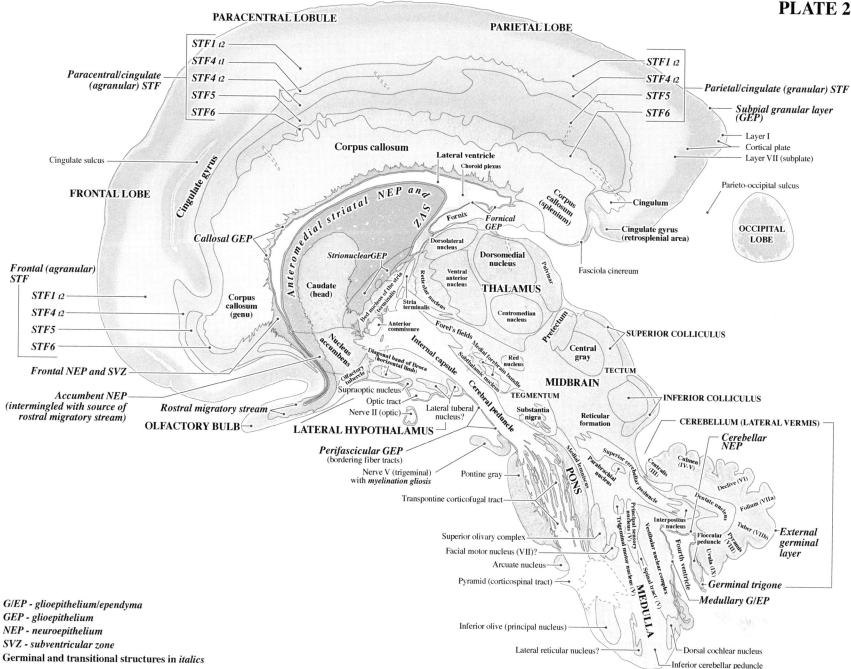

PLATE 2B

PARACENTRAL LOBULE

PARIETAL LOBE

STF1 *t2*
STF4 *t1*
STF4 *t2*
STF5
STF6

Paracentral/cingulate (agranular) STF

STF1 *t2*
STF4 *t2*
STF5
STF6

Parietal/cingulate (granular) STF

Corpus callosum

Lateral ventricle
Choroid plexus

Subpial granular layer (GEP)

Layer I
Cortical plate
Layer VII (subplate)

Cingulate sulcus

Cingulate gyrus

FRONTAL LOBE

Anteromedial striatal NEP and SVZ

Corpus callosum (splenium)

Cingulum

Parieto-occipital sulcus

Callosal GEP

Fornix
Fornical GEP

Dorsolateral nucleus

Dorsomedial nucleus

Cingulate gyrus (retrosplenial area)

OCCIPITAL LOBE

Fasciola cinereum

StrionuclearGEP

Ventral anterior nucleus

THALAMUS

Pulvinar

Frontal (agranular) STF

Caudate (head)

Reticular nucleus

Bed nucleus of the stria terminalis

Centromedian nucleus

Pretectum

SUPERIOR COLLICULUS

STF1 *t2*
STF4 *t2*
STF5
STF6

Corpus callosum (genu)

Stria terminalis

Forel's fields

Central gray

Frontal NEP and SVZ

Nucleus accumbens

Anterior commissure

Medial forebrain bundle

Subthalamic nucleus

Red nucleus

MIDBRAIN

TECTUM

INFERIOR COLLICULUS

Accumbent NEP (intermingled with source of rostral migratory stream)

Diagonal band of Broca (horizontal limb)

Internal capsule

TEGMENTUM

Olfactory tubercle

Cerebral peduncle

Substantia nigra

Reticular formation

CEREBELLUM (LATERAL VERMIS)

Rostral migratory stream

Supraoptic nucleus
Optic tract
Nerve II (optic)

Lateral tuberal nucleus?

Cerebellar NEP

OLFACTORY BULB

LATERAL HYPOTHALAMUS

Superior cerebellar peduncle

Centralis (III)

Culmen (IV-V)

Perifascicular GEP (bordering fiber tracts)

Parabrachial nucleus

Declive (VI)

Nerve V (trigeminal) with *myelination gliosis*

Pontine gray

Medial lemniscus

Dentate nucleus

Folium (VIIa)

Transpontine corticofugal tract

PONS

Principal sensory nucleus (V)

Vestibular nuclear complex

Interpositus nucleus

Tuber (VIIb)

External germinal layer

Superior olivary complex

Trigeminal motor nucleus (V)

Flocular peduncle

Pyramis (VIII)

Facial motor nucleus (VII)?

Fourth ventricle

Uvula (IX)

Arcuate nucleus

Spinal tract (V)

Germinal trigone

Pyramid (corticospinal tract)

MEDULLA

Medullary G/EP

Inferior olive (principal nucleus)

Lateral reticular nucleus?

Dorsal cochlear nucleus
Inferior cerebellar peduncle

G/EP - glioepithelium/ependyma
GEP - glioepithelium
NEP - neuroepithelium
SVZ - subventricular zone
Germinal and transitional structures in *italics*

PLATE 3A
CR 160 mm
GW 20
Y27-60
Sagittal
Section
470

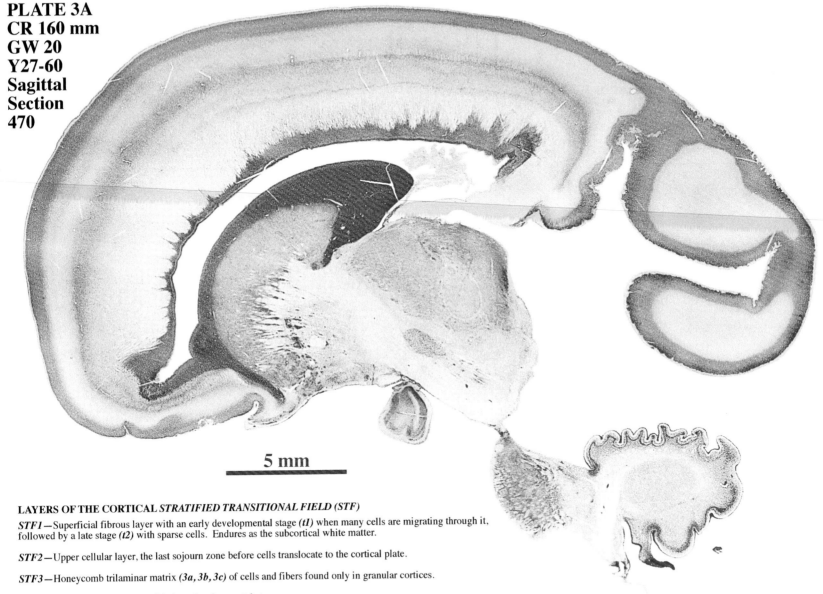

5 mm

LAYERS OF THE CORTICAL *STRATIFIED TRANSITIONAL FIELD (STF)*

STF1—Superficial fibrous layer with an early developmental stage *(t1)* when many cells are migrating through it, followed by a late stage *(t2)* with sparse cells. Endures as the subcortical white matter.

STF2—Upper cellular layer, the last sojourn zone before cells translocate to the cortical plate.

STF3—Honeycomb trilaminar matrix *(3a, 3b, 3c)* of cells and fibers found only in granular cortices.

STF4—Complex middle layer with three developmental stages:
t1– fibrous layer without interspersed cells;
t2– cells and fibers intermingle to form striations; *t3*– fibers endure in the deep white matter.

STF5—Deep cellular layer, the first sojourn zone to appear outside the germinal matrix.

STF6—Late-forming deep layer of callosal fibers outside the germinal matrix.

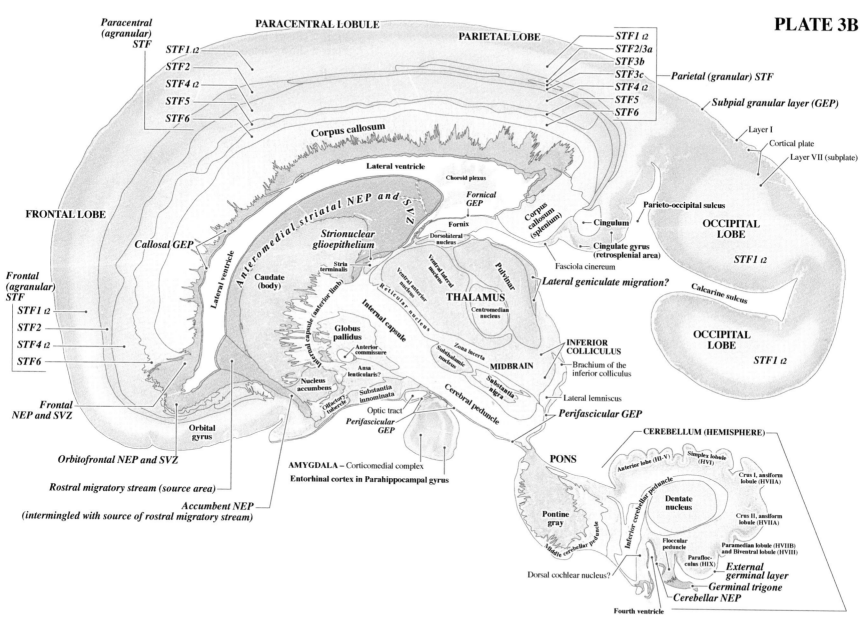

11

PLATE 3B

Paracentral (agranular) STF

STF1 t2
STF2
STF4 t2
STF5
STF6

PARACENTRAL LOBULE

PARIETAL LOBE

STF1 t2
STF2/3a
STF3b
STF3c
STF4 t2
STF5
STF6

Parietal (granular) STF

Subpial granular layer (GEP)

Layer I
Cortical plate
Layer VII (subplate)

Corpus callosum

Lateral ventricle

Choroid plexus

Fornical GEP

FRONTAL LOBE

Callosal GEP

Anteromedial striatal NEP and SVZ

Strionuclear glioepithelium

Fornix

Stria terminalis

Dorsolateral nucleus

Corpus callosum (splenium)

Cingulum

Parieto-occipital sulcus

OCCIPITAL LOBE

Fasciola cinereum

Cingulate gyrus (retrosplenial area)

STF1 t2

Lateral ventricle

Caudate (body)

Ventral anterior nucleus

Ventral lateral nucleus

Pulvinar

Lateral geniculate migration?

Calcarine sulcus

Frontal (agranular) STF

STF1 t2
STF2
STF4 t2
STF6

Reticular nucleus

THALAMUS

Centromedian nucleus

OCCIPITAL LOBE

STF1 t2

Globus pallidus

Internal capsule

Anterior commissure

Zona incerta

Subthalamic nucleus

INFERIOR COLLICULUS

MIDBRAIN

Ansa lenticularis?

Substantia nigra

Brachium of the inferior colliculus

Frontal NEP and SVZ

Nucleus accumbens

Cerebral peduncle

Lateral lemniscus

Orbital gyrus

Olfactory tubercle

Substantia innominata

Optic tract

Perifascicular GEP

Orbitofrontal NEP and SVZ

Perifascicular GEP

CEREBELLUM (HEMISPHERE)

PONS

Anterior lobe (HI-V)

Simplex lobule (HVI)

Rostral migratory stream (source area)

AMYGDALA – Corticomedial complex

Entorhinal cortex in Parahippocampal gyrus

Crus I, ansiform lobule (HVIIA)

Dentate nucleus

Accumbent NEP

(intermingled with source of rostral migratory stream)

Pontine gray

Crus II, ansiform lobule (HVIIA)

Paramedian lobule (HVIIB) and Biventral lobule (HVIII)

Floccular peduncle

Paraflocculus (HIX)

External germinal layer

Germinal trigone

Cerebellar NEP

Dorsal cochlear nucleus?

Fourth ventricle

GEP - glioepithelium
NEP - neuroepithelium
SVZ - subventricular zone
Germinal and transitional structures in italics

12

PLATE 4A
CR 160 mm
GW 20
Y27-60
Sagittal
Section
539

Plate 7 shows details of the paracentral cortex in Section 219.
Plate 8 shows details of the occipital cortex in Section 170.

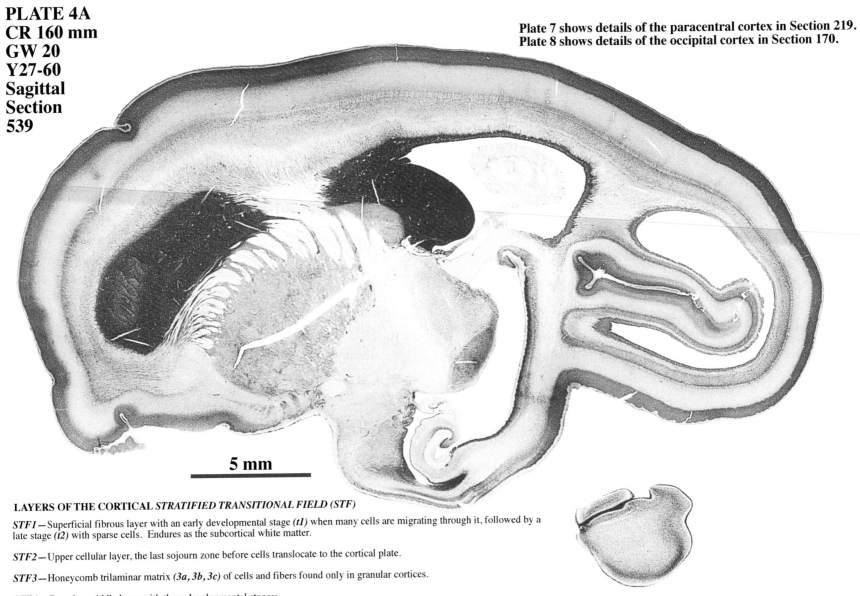

5 mm

LAYERS OF THE CORTICAL *STRATIFIED TRANSITIONAL FIELD (STF)*

STF1—Superficial fibrous layer with an early developmental stage *(t1)* when many cells are migrating through it, followed by a late stage *(t2)* with sparse cells. Endures as the subcortical white matter.

STF2—Upper cellular layer, the last sojourn zone before cells translocate to the cortical plate.

STF3—Honeycomb trilaminar matrix *(3a, 3b, 3c)* of cells and fibers found only in granular cortices.

STF4—Complex middle layer with three developmental stages:
t1– fibrous layer without interspersed cells;
t2– cells and fibers intermingle to form striations; *t3*– fibers endure in the deep white matter.

STF5—Deep cellular layer, the first sojourn zone to appear outside the germinal matrix.

STF6—Late-forming deep layer of callosal fibers outside the germinal matrix.

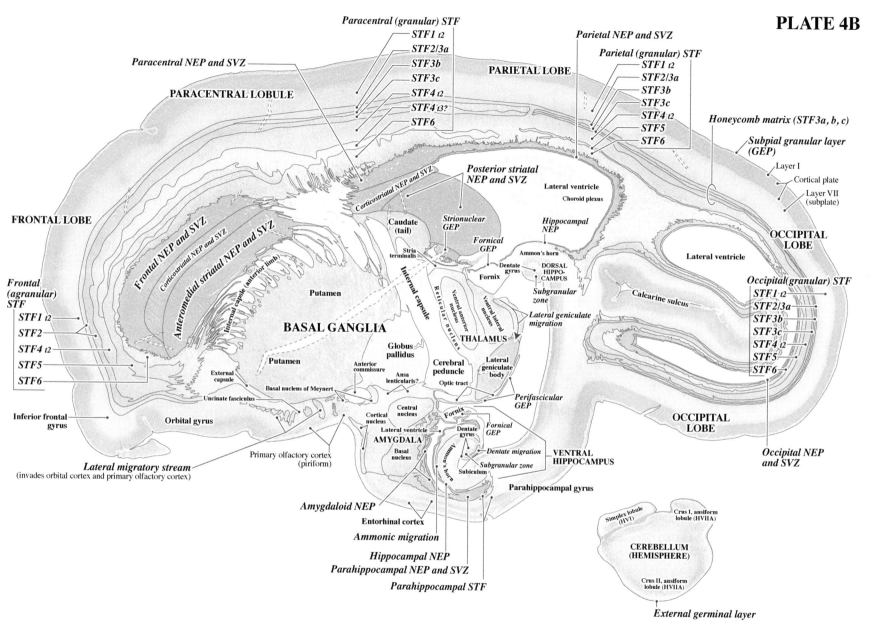

13

PLATE 4B

GEP - glioepithelium
NEP - neuroepithelium
SVZ - subventricular zone
Germinal and transitional structures in *italics*

14

PLATE 5A
CR 160 mm, GW 20, Y27-60
Sagittal, Section 590

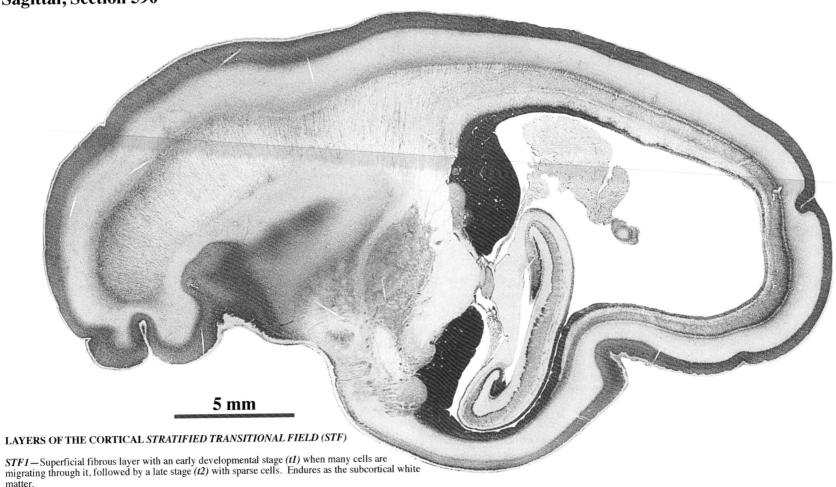

5 mm

LAYERS OF THE CORTICAL *STRATIFIED TRANSITIONAL FIELD (STF)*

STF1—Superficial fibrous layer with an early developmental stage *(t1)* when many cells are migrating through it, followed by a late stage *(t2)* with sparse cells. Endures as the subcortical white matter.

STF2—Upper cellular layer, the last sojourn zone before cells translocate to the cortical plate.

STF3—Honeycomb trilaminar matrix *(3a, 3b, 3c)* of cells and fibers found only in granular cortices.

STF4—Complex middle layer with three developmental stages:
t1– fibrous layer without interspersed cells;
t2– cells and fibers intermingle to form striations; *t3*– fibers endure in the deep white matter.

STF5—Deep cellular layer, the first sojourn zone to appear outside the germinal matrix.

STF6—Late-forming deep layer of callosal fibers outside the germinal matrix.

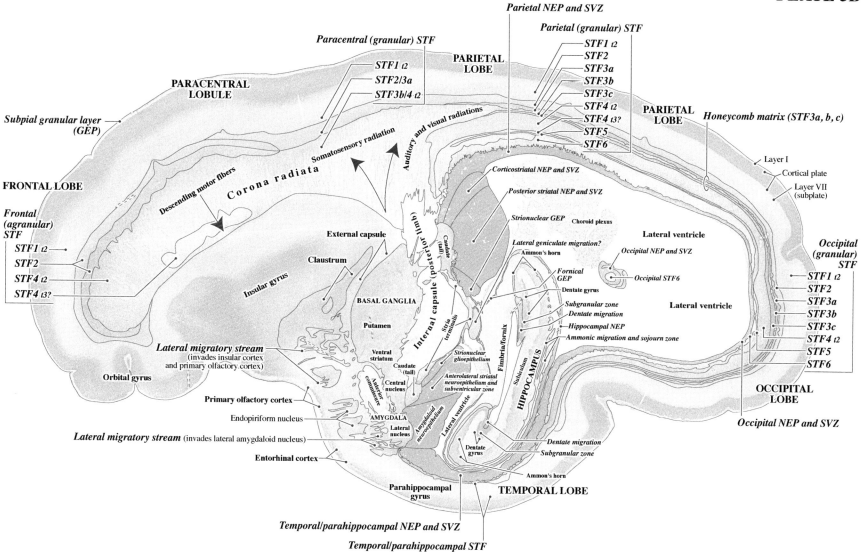

15

PLATE 5B

Parietal NEP and SVZ

Parietal (granular) STF

Paracentral (granular) STF

PARIETAL LOBE

STF1 t2
STF2
STF3a
STF3b
STF3c
STF4 t2
STF4 t3?
STF5
STF6

PARIETAL LOBE

Honeycomb matrix (STF3a, b, c)

STF1 t2
STF2/3a
STF3b/4 t2

PARACENTRAL LOBULE

Subpial granular layer (GEP)

Somatosensory radiation

Auditory and visual radiations

Layer I
Cortical plate
Layer VII (subplate)

Corticostriatal NEP and SVZ

Descending motor fibers

Corona radiata

Posterior striatal NEP and SVZ

FRONTAL LOBE

Strionuclear GEP
Choroid plexus

Lateral ventricle

Occipital (granular) STF

Frontal (agranular) STF

External capsule

Lateral geniculate migration?

Occipital NEP and SVZ

STF1 t2
STF2
STF4 t2
STF4 t3?

Claustrum

Ammon's horn

Occipital STF6

STF1 t2
STF2
STF3a
STF3b
STF3c
STF4 t2
STF5
STF6

Fornical GEP

Lateral ventricle

BASAL GANGLIA

Insular gyrus

Dentate gyrus

Subgranular zone

Dentate migration

Hippocampal NEP

Putamen

Ammonic migration and sojourn zone

Lateral migratory stream (invades insular cortex and primary olfactory cortex)

Ventral striatum

Caudate (tail)

Central nucleus

Strionuclear glioepithelium

Anterolateral striatal neuroepithelium and subventricular zone

OCCIPITAL LOBE

Orbital gyrus

Primary olfactory cortex

Endopiriform nucleus

Anterior commissure

AMYGDALA

Lateral nucleus

Lateral ventricle

Occipital NEP and SVZ

Lateral migratory stream (invades lateral amygdaloid nucleus)

Dentate migration

Subgranular zone

Entorhinal cortex

Dentate gyrus

Ammon's horn

Parahippocampal gyrus

TEMPORAL LOBE

Temporal/parahippocampal NEP and SVZ

Temporal/parahippocampal STF

GEP - glioepithelium
NEP - neuroepithelium
SVZ - subventricular zone
Germinal and transitional structures in *italics*

PLATE 6A
CR 160 mm, GW 20, Y27-60
Sagittal, Section 630

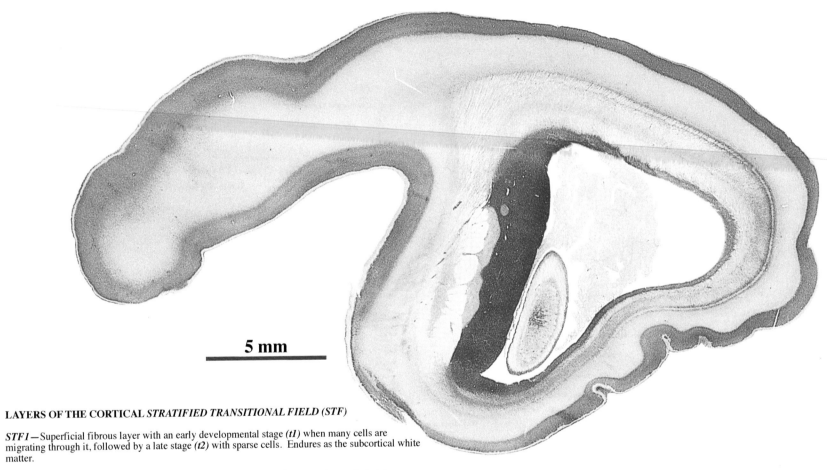

LAYERS OF THE CORTICAL *STRATIFIED TRANSITIONAL FIELD (STF)*

STF1—Superficial fibrous layer with an early developmental stage *(t1)* when many cells are
migrating through it, followed by a late stage *(t2)* with sparse cells. Endures as the subcortical white
matter.

STF2—Upper cellular layer, the last sojourn zone before cells translocate to the cortical plate.

STF3—Honeycomb trilaminar matrix *(3a, 3b, 3c)* of cells and fibers found only in granular cortices.

STF4—Complex middle layer with three developmental stages:
t1– fibrous layer without interspersed cells;
t2– cells and fibers intermingle to form striations; *t3*– fibers endure in the deep white matter.

STF5—Deep cellular layer, the first sojourn zone to appear outside the germinal matrix.

STF6—Late-forming deep layer of callosal fibers outside the germinal matrix.

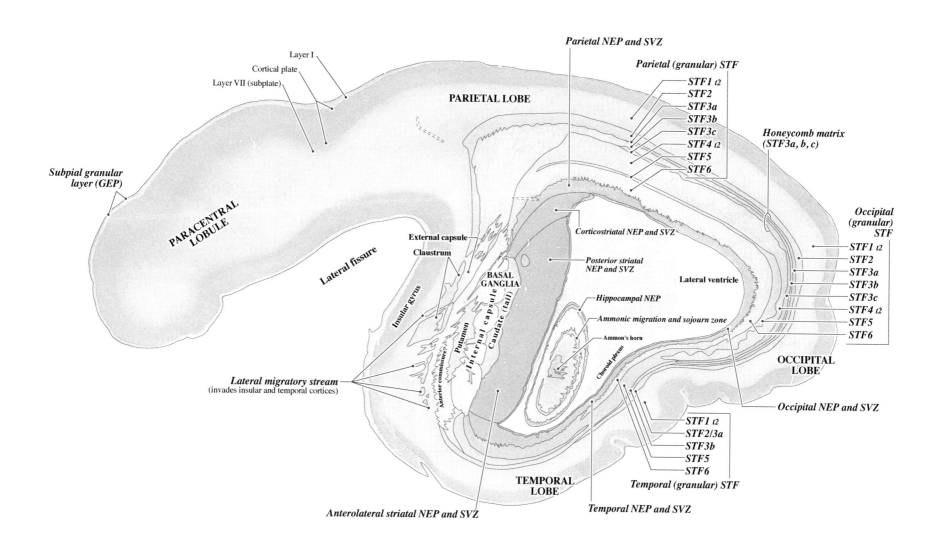

Parietal NEP and SVZ

PARIETAL LOBE

Parietal (granular) STF

STF1 t2
STF2
STF3a
STF3b
STF3c
STF4 t2
STF5
STF6

Honeycomb matrix (STF3a, b, c)

Layer I

Cortical plate

Layer VII (subplate)

Subpial granular layer (GEP)

PARACENTRAL LOBULE

Lateral fissure

External capsule

Claustrum

Corticostriatal NEP and SVZ

Posterior striatal NEP and SVZ

BASAL GANGLIA

Insular gyrus

Putamen
Internal capsule
Caudate (tail)
Anterior commissure?

Hippocampal NEP

Ammonic migration and sojourn zone

Ammon's horn

Choroid plexus

Lateral ventricle

Occipital (granular) STF

STF1 t2
STF2
STF3a
STF3b
STF3c
STF4 t2
STF5
STF6

OCCIPITAL LOBE

Occipital NEP and SVZ

Lateral migratory stream
(invades insular and temporal cortices)

STF1 t2
STF2/3a
STF3b
STF5
STF6

TEMPORAL LOBE

Temporal (granular) STF

Anterolateral striatal NEP and SVZ

Temporal NEP and SVZ

GEP - glioepithelium
NEP - neuroepithelium
SVZ - subventricular zone
Germinal and transitional structures in *italics*

PLATE 7
CR 160 mm, GW 20, Y27-60, Sagittal, Section 219
PARACENTRAL CORTEX

A. A low-magnification view of Section 219 (*compare* with **Plate 4, Section 539** from the opposite side).

B. A slice of the paracentral cortex. There are substantial thickness differences between the *stratified transitional field (STF)*, the *cortical plate*, and the combined *neuroepithelium and subventricular zone (NEP+SVZ)*. Actual thicknesses are:
NEP+SVZ=0.563 mm,
Cortical plate=0.615 mm,
STF=3.54 mm
The *STF* is 6.29 times thicker than the *NEP+SVZ* and 5.76 times thicker than the cortical plate.

The *rectangles* show the areas enlarged in **C and D**. This is a cortical area where Layer V will be more prominent than Layer IV in the mature (agranular) cortex.

C. The paracentral cortical plate. *Roman numerals* indicate the presumptive cortical layers. Cells are densely packed, especially superficially in what is presumably Layers **II** and **III**. Layer **IV** is indistinct, but Layers **V** and **VI** are thick.

D. The paracentral *STF, NEP,* and *SVZ.* Note that *STF3* is absent, typical of agranular cortices.

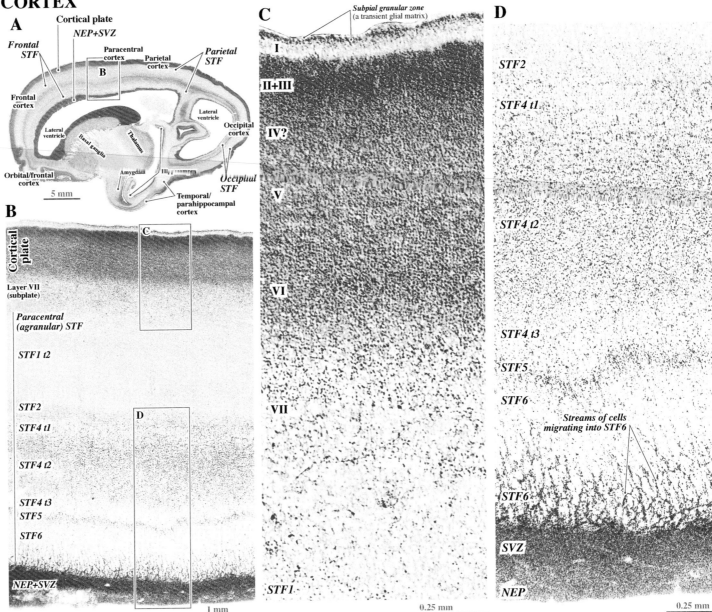

A

Cortical plate
NEP+SVZ

Frontal STF

Paracentral cortex

Parietal cortex

Parietal STF

B

Frontal cortex

Lateral ventricle

Lateral ventricle

Basal ganglia

Thalamus

Orbital/frontal cortex

Amygdala

III ventricle

Temporal/parahippocampal cortex

Occipital cortex

Occipital STF

5 mm

B

Cortical plate

Layer VII (subplate)

Paracentral (agranular) STF

STF1 t2

STF2

STF4 t1

STF4 t2

STF4 t3
STF5

STF6

NEP+SVZ

C

D

1 mm

C

Subpial granular zone (a transient glial matrix)

I

II+III

IV?

V

VI

VII

STF1

0.25 mm

D

STF2

STF4 t1

STF4 t2

STF4 t3

STF5

STF6

Streams of cells migrating into STF6

STF6

SVZ

NEP

0.25 mm

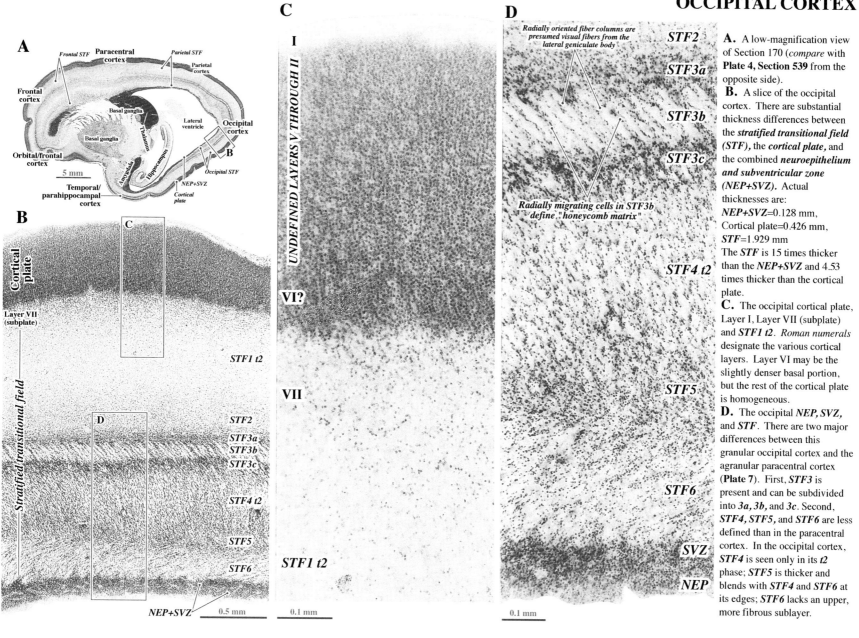

PLATE 8
CR 160 mm, GW 20, Y27-60, Sagittal, Section 170
OCCIPITAL CORTEX

A. A low-magnification view of Section 170 (*compare* with **Plate 4, Section 539** from the opposite side).

B. A slice of the occipital cortex. There are substantial thickness differences between the *stratified transitional field (STF)*, the *cortical plate*, and the combined *neuroepithelium and subventricular zone (NEP+SVZ)*. Actual thicknesses are: NEP+SVZ=0.128 mm, Cortical plate=0.426 mm, *STF*=1.929 mm The *STF* is 15 times thicker than the *NEP+SVZ* and 4.53 times thicker than the cortical plate.

C. The occipital cortical plate, Layer I, Layer VII (subplate) and *STF1 t2*. Roman numerals designate the various cortical layers. Layer VI may be the slightly denser basal portion, but the rest of the cortical plate is homogeneous.

D. The occipital *NEP, SVZ,* and *STF*. There are two major differences between this granular occipital cortex and the agranular paracentral cortex (**Plate 7**). First, *STF3* is present and can be subdivided into *3a, 3b,* and *3c*. Second, *STF4, STF5,* and *STF6* are less defined than in the paracentral cortex. In the occipital cortex, *STF4* is seen only in its *t2* phase; *STF5* is thicker and blends with *STF4* and *STF6* at its edges; *STF6* lacks an upper, more fibrous sublayer.

PLATE 9
CR 160 mm, GW 20, Y27-60, Sagittal, Section 379
CEREBELLUM, CULMEN

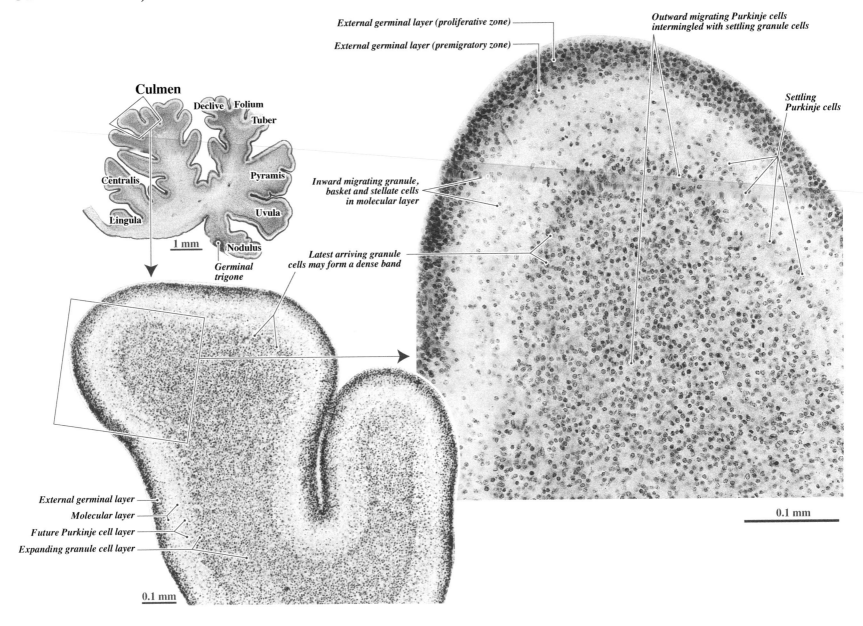

Culmen

Declive   Folium
Tuber

Centralis

Pyramis

Lingula

Uvula

Nodulus

1 mm

Germinal
trigone

External germinal layer (proliferative zone)

External germinal layer (premigratory zone)

Outward migrating Purkinje cells
intermingled with settling granule cells

Settling
Purkinje cells

Inward migrating granule,
basket and stellate cells
in molecular layer

Latest arriving granule
cells may form a dense band

External germinal layer
Molecular layer
Future Purkinje cell layer
Expanding granule cell layer

0.1 mm

0.1 mm

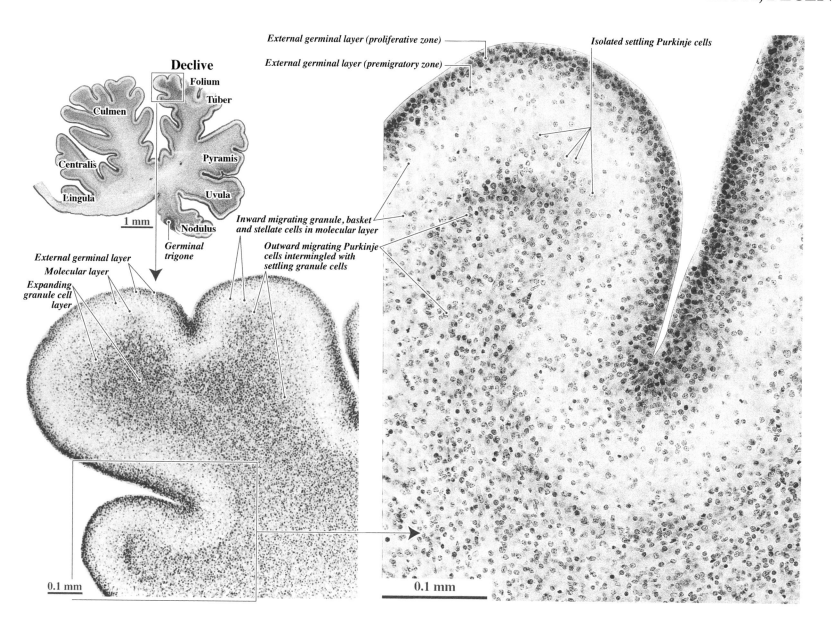

Declive

Culmen

Folium

Tuber

Centralis

Pyramis

Lingula

Uvula

Nodulus

Germinal trigone

1 mm

External germinal layer (proliferative zone)

External germinal layer (premigratory zone)

Isolated settling Purkinje cells

External germinal layer

Molecular layer

Expanding granule cell layer

Inward migrating granule, basket and stellate cells in molecular layer

Outward migrating Purkinje cells intermingled with settling granule cells

0.1 mm

0.1 mm

PART III: Y17-63
CR 165 mm (GW 20)
Frontal

This specimen is a stillborn fetus of undetermined sex (Yakovlev case number BR-17-63, our Y17-63) with a crown-rump length (CR) of 165-mm estimated to be at gestational week (GW) 20. The brain was cut in the frontal plane in 1,002 sections (35-μm thick) and is classified as a Normative Control in the Yakovlev Collection (Haleem, 1990). Since there is no photograph of this brain before it was embedded and cut, we turned to the comprehensive treatise that Retzius published in 1896 showing whole fetal brains in medial, lateral, superior, and inferior views and midline sagittally cut brains. **Figure 4**, taken from Retzius (1896), shows the exterior of a brain from a specimen that is at the same age as Y17-63, along with the approximate cutting angle of the sections. Photographs of 12 different Nissl-stained sections are illustrated in **Plates 11–22**. **Plates 23–34** show various high magnification views of the brainstem. Different areas of the cerebral cortex are shown at still higher magnification in **Plates 35-38**. High-magnification views of the crossed and uncrossed corpus callosum are shown in **Plates 39-40**.

In the cortical regions of the telencephalon, a thick *neuroepithelium/subventricular zone* is generating neocortical neurons mainly for Layers II, III, and IV. Many neurons and/or glia are sojourning in more definite *stratified transitional field* layers in all lobes of the cerebral cortex. There is a pronounced regional heterogeneity between granular (future sensory) and agranular (future motor) areas. In anterolateral parts of the cerebral cortex, streams of neurons and glia are numerous in the *lateral migratory stream*. That stream percolates through the claustrum, endopiriform nucleus, external capsule, and uncinate fasciculus, and the cells appear to be heading toward the insular cortex, primary olfactory cortex, temporal cortex, and basolateral parts of the amygdaloid complex. In the hippocampus, cells are entering Ammon's horn pyramidal layer via a more definite *ammonic migratory stream*, and granule cells and their precursors are migrating to the dentate gyrus in the *dentate migratory stream*. Stem cells of dentate granule cells populate the periphery of the dentate hilus just beneath the granular layer to form the *subgranular zone* where dentate granule cells are being generated. A large *neuroepithelium/subventricular zone* overlies the nucleus accumbens and striatum where neurons (and glia) are being generated. But there are less distinct subdivisions between anterolateral, anteromedial, and posterior parts of the *striatal neuroepithelium and subventricular zone*. The septum has a *glioepithelium/ependyma* at the ventricular surface. More

neurons, glia, and their mitotic precursor cells are migrating through the olfactory peduncle toward the olfactory bulb (*rostral migratory stream*). Neurons in most diencephalic structures appear to be settled and are maturing; the third ventricle is lined with a thin *glioepithelium/ependyma*. There is no lamination in the lateral geniculate body. In the midbrain, pons, and medulla, a *glioepithelium/ependyma* lines the cerebral aqueduct and fourth ventricle.

The cerebellum is continuing its growth. The dentate nucleus is not laminated in the deep white matter. The surface of the cerebellar cortex is covered by the *external germinal layer (egl)* that is actively growing and producing some basket, stellate, and granule cells. A thin molecular layer is beneath the *egl*. Below that is an indefinite Purkinje cell layer (most Purkinje cells are still migrating). The granule cell layer is broad and diffuse in some areas, especially in the posterolateral hemisphere, but in other areas of the hemisphere and in the vermis, there is a dense superficial part, possibly composed of newly arrived migrating neurons. The vermal and hemispheric lobules are broad and show little secondary infolding. But the *germinal trigone* at the base of the nodulus and along the floccular peduncle is prominent.

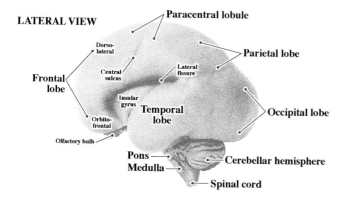

LATERAL VIEW

Paracentral lobule

Dorso-lateral

Central sulcus

Lateral fissure

Parietal lobe

Frontal lobe

Insular gyrus

Temporal lobe

Occipital lobe

Orbito-frontal

Olfactory bulb

Pons

Medulla

Cerebellar hemisphere

Spinal cord

SUPERIOR VIEW

Interhemispheric fissure

Frontal lobe (dorsal)

Paracentral lobule

Lateral fissure

Temporal lobe

Parietal lobe

Occipital lobe

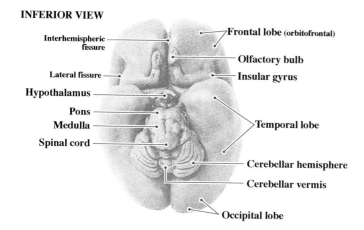

INFERIOR VIEW

Interhemispheric fissure

Frontal lobe (orbitofrontal)

Olfactory bulb

Lateral fissure

Insular gyrus

Hypothalamus

Pons

Medulla

Temporal lobe

Spinal cord

Cerebellar hemisphere

Cerebellar vermis

Occipital lobe

GW20 FRONTAL SECTION PLANES

SECTION 281   361   441   481   511   551   601   631   661   701   741   821

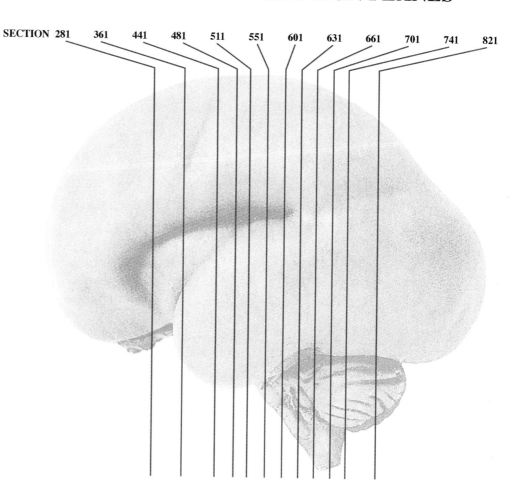

Figure 4. The column on the *left* shows a GW20 brain from lateral, superior, and inferior views with major structures labeled (Figures 1, 2, and 4 in Table 9, Volume 2, Retzius, 1896). An enlarged lateral view with the approximate locations and cutting angles of the sections of Y17-63 is shown *above*.

24

PLATE 11A
CR 165 mm, GW 20, Y17-63
Frontal, Section 281

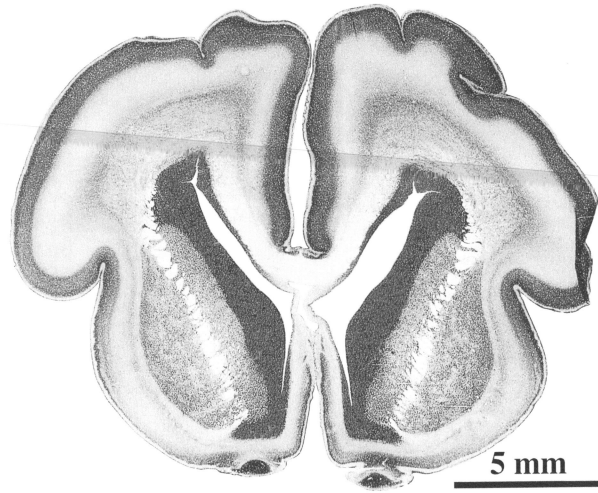

5 mm

LAYERS OF THE CORTICAL
STRATIFIED TRANSITIONAL
FIELD (STF)

STF1—Superficial fibrous
layer with an early
developmental stage *(t1)*
when many cells are
migrating through it, followed
by a late stage *(t2)* with sparse
cells. Endures as the
subcortical white matter.

STF2—Upper cellular layer,
the last sojourn zone before
cells translocate to the cortical
plate.

STF3—Honeycomb
trilaminar matrix *(3a, 3b, 3c)*
of cells and fibers found only
in granular cortices.

STF4—Complex middle
layer with three
developmental stages:
t1– fibrous layer without
interspersed cells;
t2– cells and fibers
intermingle to form striations;
t3– fibers endure in the deep
white matter.

STF5—Deep cellular layer,
the first sojourn zone to
appear outside the germinal
matrix.

STF6—Late-forming deep
layer of callosal fibers outside
the germinal matrix.

See detail of the brain core in Plates 23A and B.

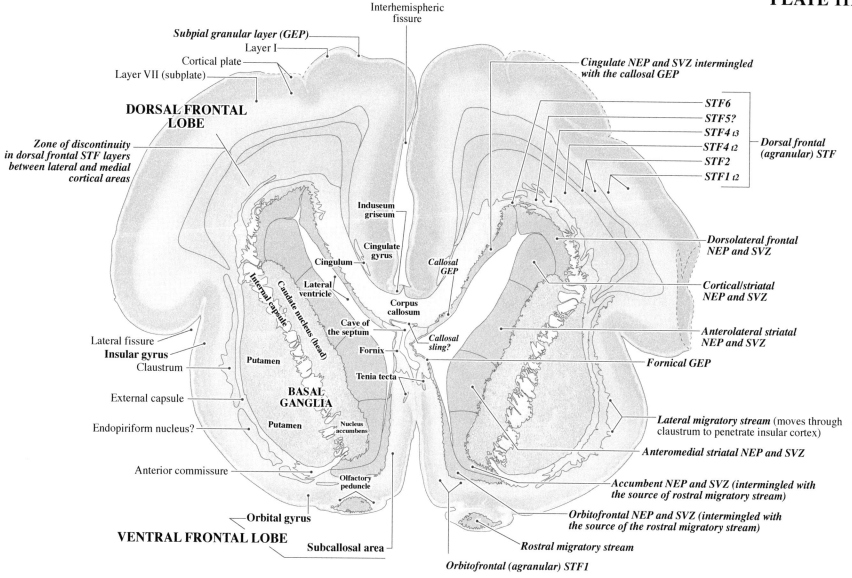

Interhemispheric fissure

Subpial granular layer (GEP)
Layer I
Cortical plate
Layer VII (subplate)

Cingulate NEP and SVZ intermingled with the callosal GEP

DORSAL FRONTAL LOBE

STF6
STF5?
STF4 *t3*
STF4 *t2*
STF2
STF1 *t2*

} *Dorsal frontal (agranular) STF*

Zone of discontinuity in dorsal frontal STF layers between lateral and medial cortical areas

Induseum griseum

Cingulate gyrus

Dorsolateral frontal NEP and SVZ

Cingulum

Callosal GEP

Cortical/striatal NEP and SVZ

Lateral ventricle

Internal capsule

Caudate nucleus (head)

Corpus callosum

Cave of the septum

Callosal sling?

Anterolateral striatal NEP and SVZ

Lateral fissure
Insular gyrus
Claustrum

Fornix

Tenia tecta

Fornical GEP

External capsule

BASAL GANGLIA

Putamen

Nucleus accumbens

Lateral migratory stream (moves through claustrum to penetrate insular cortex)

Endopiriform nucleus?

Putamen

Anteromedial striatal NEP and SVZ

Anterior commissure

Olfactory peduncle

Accumbent NEP and SVZ (intermingled with the source of rostral migratory stream)

Orbitofrontal NEP and SVZ (intermingled with the source of the rostral migratory stream)

Orbital gyrus

VENTRAL FRONTAL LOBE

Subcallosal area

Rostral migratory stream

Orbitofrontal (agranular) STF1

GEP - *glioepithelium*
NEP - *neuroepithelium*
SVZ - *subventricular zone*
Germinal and transitional structures in *italics*

PLATE 12A
CR 165 mm, GW 20, Y17-63
Frontal, Section 361

See this area of cortex from
Section 341 in Plate 35.

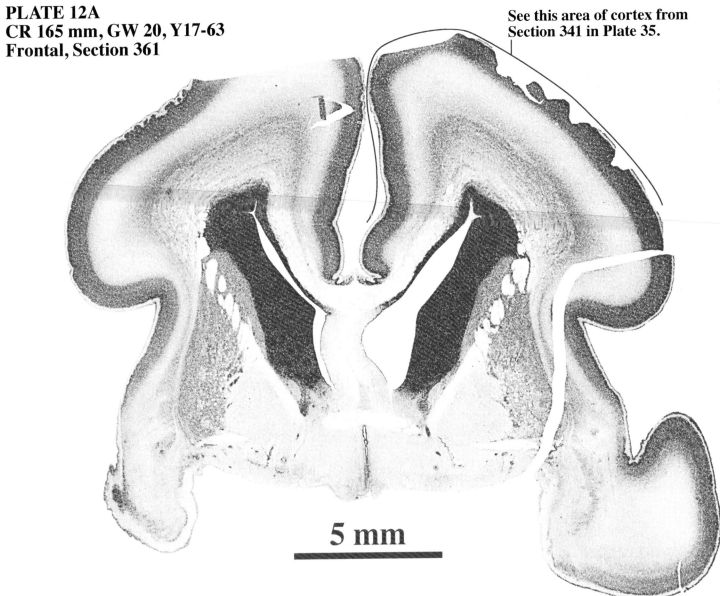

5 mm

LAYERS OF THE CORTICAL
STRATIFIED TRANSITIONAL
FIELD (STF)

STF1—Superficial fibrous
layer with an early
developmental stage *(t1)*
when many cells are
migrating through it, followed
by a late stage *(t2)* with sparse
cells. Endures as the
subcortical white matter.

STF2—Upper cellular layer,
the last sojourn zone before
cells translocate to the cortical
plate.

STF3—Honeycomb
trilaminar matrix *(3a, 3b, 3c)*
of cells and fibers found only
in granular cortices.

STF4—Complex middle
layer with three
developmental stages:
t1– fibrous layer without
interspersed cells;
t2– cells and fibers
intermingle to form striations;
t3– fibers endure in the deep
white matter.

STF5—Deep cellular layer,
the first sojourn zone to
appear outside the germinal
matrix.

STF6—Late-forming deep
layer of callosal fibers outside
the germinal matrix.

See detail of the brain core in Plates 24A and B.
A high-magnification view of the corpus callosum from Section 381 is in Plate 39.

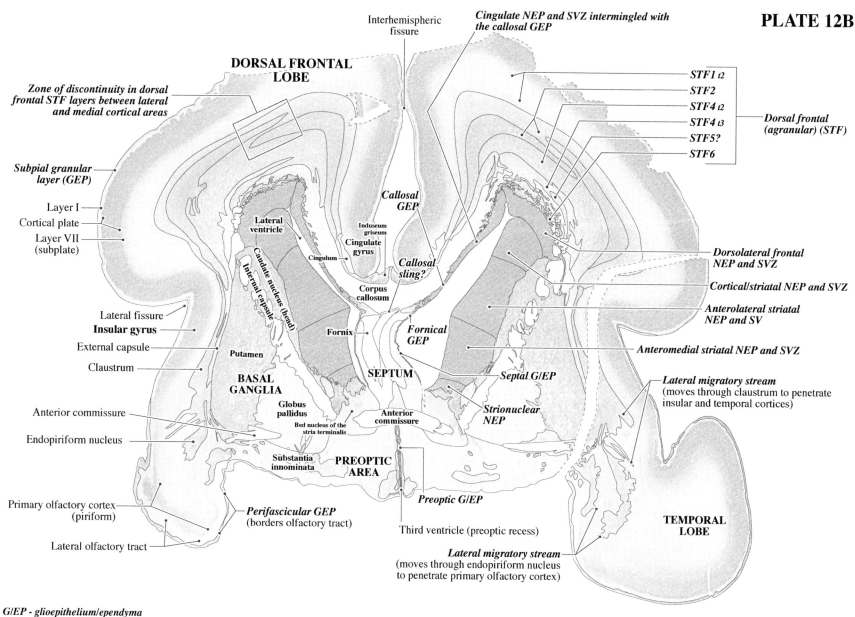

27

PLATE 12B

Interhemispheric fissure

Cingulate NEP and SVZ intermingled with the callosal GEP

DORSAL FRONTAL LOBE

Zone of discontinuity in dorsal frontal STF layers between lateral and medial cortical areas

STF1 t2
STF2
STF4 t2
STF4 t3
STF5?
STF6

Dorsal frontal (agranular) (STF)

Subpial granular layer (GEP)

Callosal GEP

Layer I

Cortical plate

Layer VII (subplate)

Lateral ventricle

Induseum griseum

Cingulate gyrus

Cingulum

Caudate nucleus (head)

Internal capsule

Callosal sling?

Corpus callosum

Dorsolateral frontal NEP and SVZ

Cortical/striatal NEP and SVZ

Anterolateral striatal NEP and SV

Lateral fissure

Insular gyrus

External capsule

Claustrum

Fornix

SEPTUM

Fornical GEP

Anteromedial striatal NEP and SVZ

Putamen

BASAL GANGLIA

Septal G/EP

Lateral migratory stream (moves through claustrum to penetrate insular and temporal cortices)

Anterior commissure

Globus pallidus

Bed nucleus of the stria terminalis

Anterior commissure

Strionuclear NEP

Endopiriform nucleus

Substantia innominata

PREOPTIC AREA

Primary olfactory cortex (piriform)

Perifascicular GEP (borders olfactory tract)

Preoptic G/EP

TEMPORAL LOBE

Lateral olfactory tract

Third ventricle (preoptic recess)

Lateral migratory stream (moves through endopiriform nucleus to penetrate primary olfactory cortex)

G/EP - glioepithelium/ependyma
GEP - glioepithelium
NEP - neuroepithelium
SVZ - subventricular zone
Germinal and transitional structures in *italics*

**PLATE 13A
CR 165 mm, GW 20, Y17-63
Frontal, Section 441**

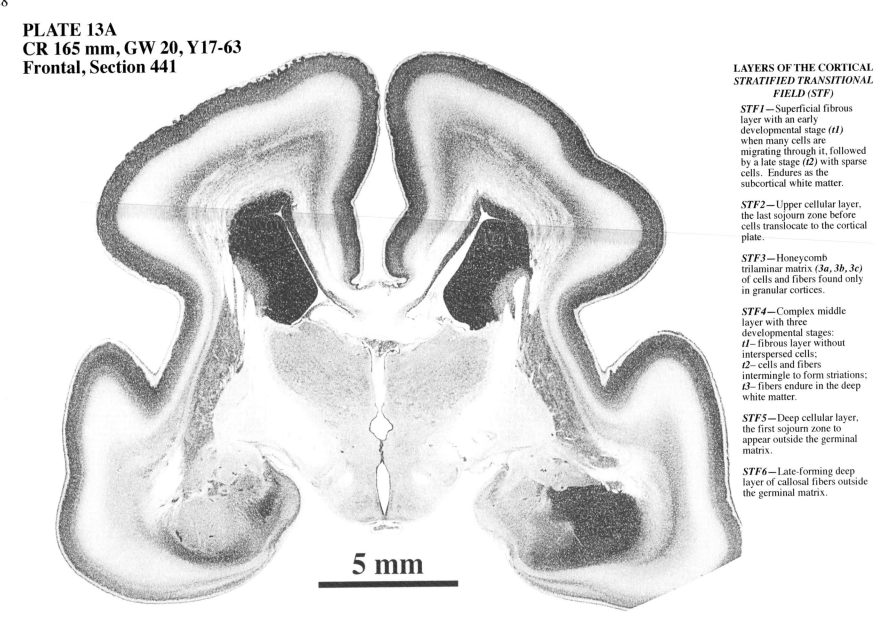

5 mm

**LAYERS OF THE CORTICAL
*STRATIFIED TRANSITIONAL
FIELD (STF)***

STF1—Superficial fibrous
layer with an early
developmental stage *(t1)*
when many cells are
migrating through it, followed
by a late stage *(t2)* with sparse
cells. Endures as the
subcortical white matter.

STF2—Upper cellular layer,
the last sojourn zone before
cells translocate to the cortical
plate.

STF3—Honeycomb
trilaminar matrix *(3a, 3b, 3c)*
of cells and fibers found only
in granular cortices.

STF4—Complex middle
layer with three
developmental stages:
t1– fibrous layer without
interspersed cells;
t2– cells and fibers
intermingle to form striations;
t3– fibers endure in the deep
white matter.

STF5—Deep cellular layer,
the first sojourn zone to
appear outside the germinal
matrix.

STF6—Late-forming deep
layer of callosal fibers outside
the germinal matrix.

**See detail of the brain core in Plates 25A and B.
A high-magnification view of the corpus callosum from Section 431 is in Plates 39 and 40.**

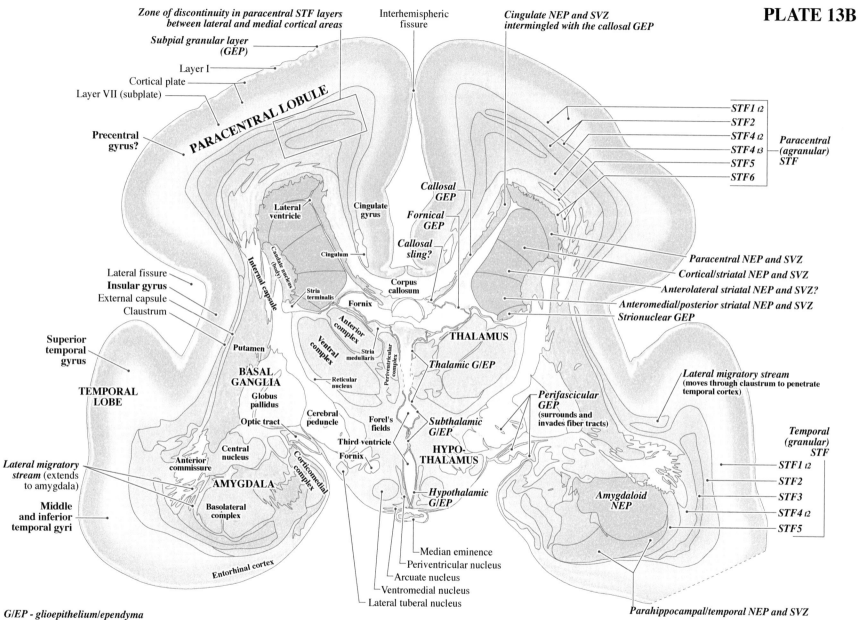

Zone of discontinuity in paracentral STF layers between lateral and medial cortical areas

Subpial granular layer (GEP)

Layer I

Cortical plate

Layer VII (subplate)

Precentral gyrus?

PARACENTRAL LOBULE

Interhemispheric fissure

Cingulate NEP and SVZ intermingled with the callosal GEP

STF1 t2
STF2
STF4 t2
STF4 t3
STF5
STF6

Paracentral (agranular) STF

Lateral ventricle

Cingulate gyrus

Cingulum

Callosal GEP

Fornical GEP

Callosal sling?

Corpus callosum

Candate nucleus (body)

Stria terminalis

Lateral fissure
Insular gyrus
External capsule
Claustrum

Internal capsule

Fornix

Anterior complex

Stria medullaris

Ventral complex

THALAMUS

Periventricular complex

Thalamic G/EP

Paracentral NEP and SVZ

Cortical/striatal NEP and SVZ

Anterolateral striatal NEP and SVZ?

Anteromedial/posterior striatal NEP and SVZ

Strionuclear GEP

Superior temporal gyrus

Putamen

Reticular nucleus

BASAL GANGLIA

Globus pallidus

Optic tract

Cerebral peduncle

Forel's fields

Subthalamic G/EP

HYPO-THALAMUS

Perifascicular GEP (surrounds and invades fiber tracts)

Lateral migratory stream (moves through claustrum to penetrate temporal cortex)

TEMPORAL LOBE

Central nucleus

Third ventricle

Fornix

Temporal (granular) STF

STF1 t2
STF2
STF3
STF4 t2
STF5

Lateral migratory stream (extends to amygdala)

Middle and inferior temporal gyri

Anterior commissure

AMYGDALA

Corticomedial complex

Basolateral complex

Hypothalamic G/EP

Amygdaloid NEP

Entorhinal cortex

Median eminence
Periventricular nucleus
Arcuate nucleus
Ventromedial nucleus
Lateral tuberal nucleus

Parahippocampal/temporal NEP and SVZ

G/EP - glioepithelium/ependyma
GEP - glioepithelium
NEP - neuroepithelium
SVZ - subventricular zone
Germinal and transitional structures in *italics*

**PLATE 14A
CR 165 mm, GW 20, Y17-63
Frontal, Section 481**

See this area of cortex from
Section 471 in Plate 37.

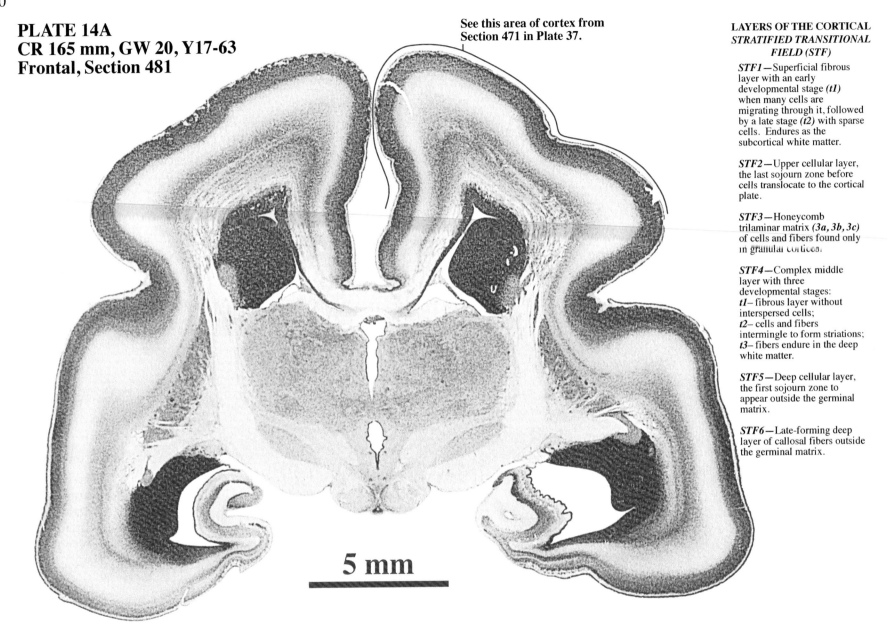

5 mm

**LAYERS OF THE CORTICAL
*STRATIFIED TRANSITIONAL
FIELD (STF)***

STF1—Superficial fibrous
layer with an early
developmental stage *(t1)*
when many cells are
migrating through it, followed
by a late stage *(t2)* with sparse
cells. Endures as the
subcortical white matter.

STF2—Upper cellular layer,
the last sojourn zone before
cells translocate to the cortical
plate.

STF3—Honeycomb
trilaminar matrix *(3a, 3b, 3c)*
of cells and fibers found only
in granular cortices.

STF4—Complex middle
layer with three
developmental stages:
t1– fibrous layer without
interspersed cells;
t2– cells and fibers
intermingle to form striations;
t3– fibers endure in the deep
white matter.

STF5—Deep cellular layer,
the first sojourn zone to
appear outside the germinal
matrix.

STF6—Late-forming deep
layer of callosal fibers outside
the germinal matrix.

See detail of the brain core in Plates 26A and B.

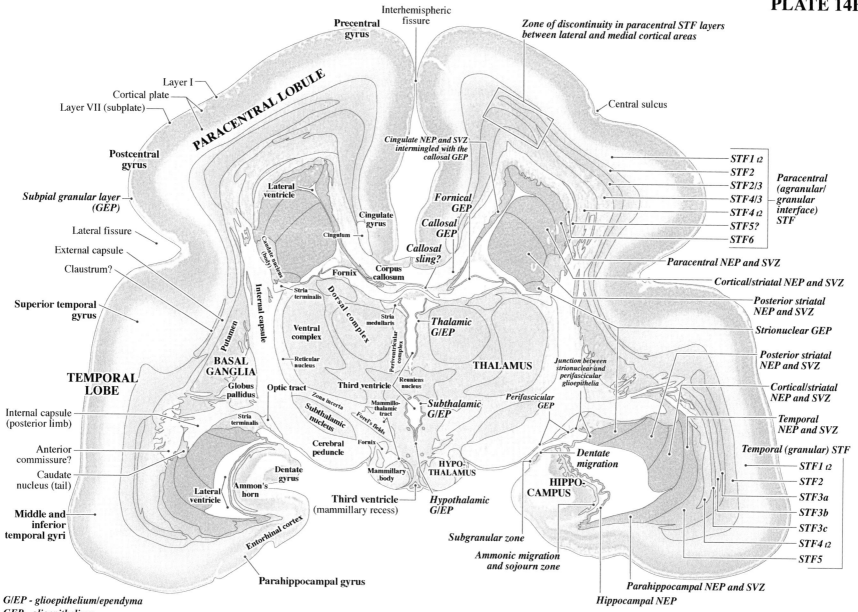

Interhemispheric fissure

Precentral gyrus

Zone of discontinuity in paracentral STF layers between lateral and medial cortical areas

Central sulcus

Layer I

Cortical plate

Layer VII (subplate)

PARACENTRAL LOBULE

Cingulate NEP and SVZ intermingled with the callosal GEP

STF1 t2
STF2
STF2/3
STF4/3
STF4 t2
STF5?
STF6

Paracentral (agranular/ granular interface) STF

Postcentral gyrus

Lateral ventricle

Fornical GEP

Subpial granular layer (GEP)

Cingulate gyrus

Callosal GEP

Paracentral NEP and SVZ

Lateral fissure

Cingulum

Cingulum

Callosal sling?

Cortical/striatal NEP and SVZ

External capsule

Caudate nucleus (body)

Corpus callosum

Posterior striatal NEP and SVZ

Claustrum?

Fornix

Strionuclear GEP

Superior temporal gyrus

Stria terminalis

Dorsal complex

Stria medullaris

Thalamic G/EP

Posterior striatal NEP and SVZ

Ventral complex

Periventricular complex

Internal capsule

Reticular nucleus

THALAMUS

Junction between strionuclear and perifascicular glioepithelia

Cortical/striatal NEP and SVZ

BASAL GANGLIA

Putamen

Reuniens nucleus

Globus pallidus

Optic tract

Third ventricle

Perifascicular GEP

Temporal NEP and SVZ

TEMPORAL LOBE

Zona incerta

Mammillo-thalamic tract

Subthalamic G/EP

Posterior striatal NEP and SVZ

Internal capsule (posterior limb)

Stria terminalis

Subthalamic nucleus

Forel's fields

Dentate migration

Temporal (granular) STF

Anterior commissure?

Cerebral peduncle

Fornix

HYPO-THALAMUS

HIPPO-CAMPUS

STF1 t2

Caudate nucleus (tail)

Dentate gyrus

Mammillary body

STF2

Middle and inferior temporal gyri

Lateral ventricle

Ammon's horn

Third ventricle (mammillary recess)

Hypothalamic G/EP

STF3a

STF3b

STF3c

Subgranular zone

STF4 t2

Entorhinal cortex

Ammonic migration and sojourn zone

STF5

Parahippocampal gyrus

Parahippocampal NEP and SVZ

Hippocampal NEP

G/EP - glioepithelium/ependyma
GEP - glioepithelium
NEP - neuroepithelium
SVZ - subventricular zone
Germinal and transitional structures in *italics*

PLATE 15A
CR 165 mm, GW 20, Y17-63
Frontal, Section 511

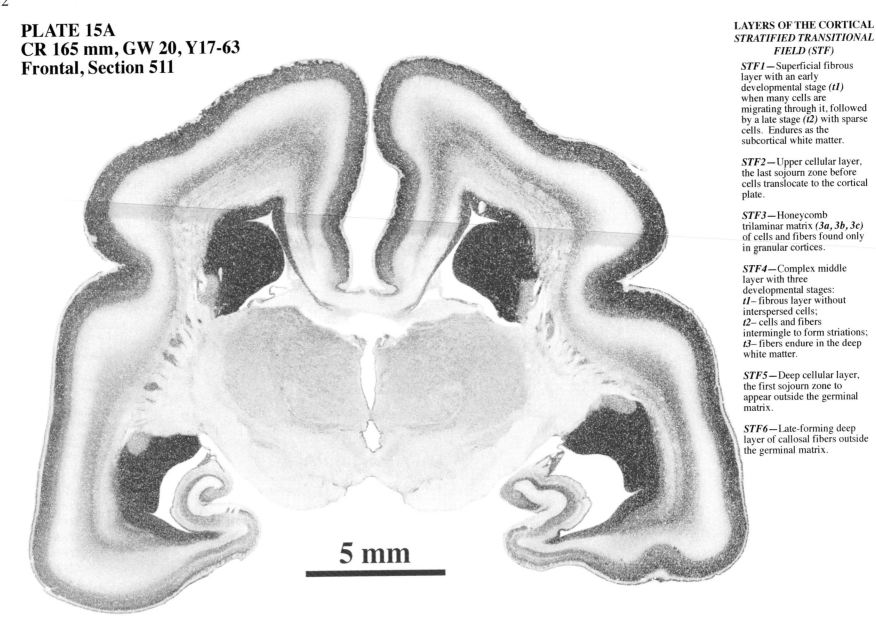

5 mm

See detail of the brain core in Plates 27A and B.

LAYERS OF THE CORTICAL
STRATIFIED TRANSITIONAL
FIELD (STF)

STF1—Superficial fibrous layer with an early developmental stage *(t1)* when many cells are migrating through it, followed by a late stage *(t2)* with sparse cells. Endures as the subcortical white matter.

STF2—Upper cellular layer, the last sojourn zone before cells translocate to the cortical plate.

STF3—Honeycomb trilaminar matrix *(3a, 3b, 3c)* of cells and fibers found only in granular cortices.

STF4—Complex middle layer with three developmental stages: *t1*– fibrous layer without interspersed cells; *t2*– cells and fibers intermingle to form striations; *t3*– fibers endure in the deep white matter.

STF5—Deep cellular layer, the first sojourn zone to appear outside the germinal matrix.

STF6—Late-forming deep layer of callosal fibers outside the germinal matrix.

PLATE 15B

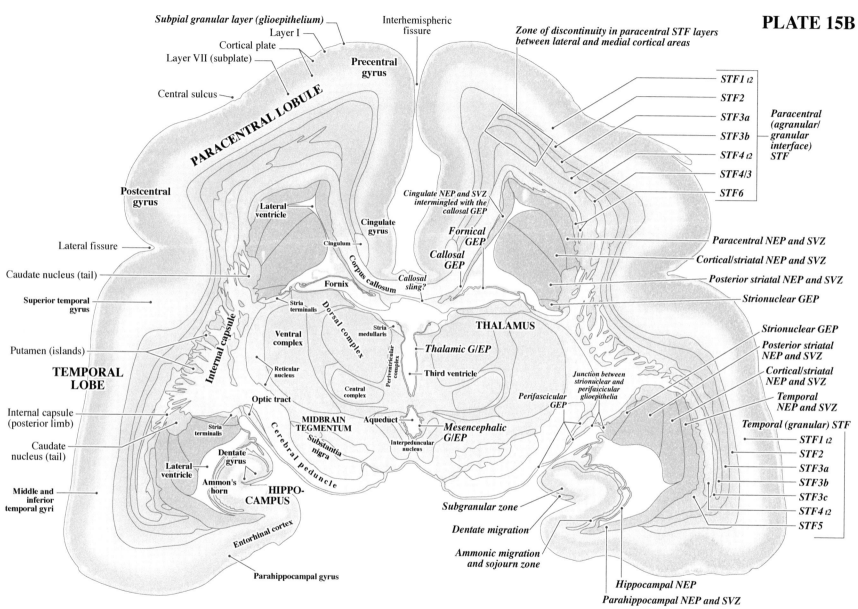

Subpial granular layer (glioepithelium)
Layer I
Cortical plate
Layer VII (subplate)
Central sulcus
Interhemispheric fissure
Precentral gyrus
PARACENTRAL LOBULE

Zone of discontinuity in paracentral STF layers between lateral and medial cortical areas

STF1 t2
STF2
STF3a
STF3b
STF4 t2
STF4/3
STF6

Paracentral (agranular/ granular interface) STF

Postcentral gyrus
Lateral ventricle
Cingulate gyrus
Cingulum
Lateral fissure
Cingulate NEP and SVZ intermingled with the callosal GEP
Fornical GEP
Callosal GEP

Paracentral NEP and SVZ
Cortical/striatal NEP and SVZ
Posterior striatal NEP and SVZ
Strionuclear GEP

Caudate nucleus (tail)
Corpus callosum
Fornix
Callosal sling?

Superior temporal gyrus
Stria terminalis
Dorsal complex
Stria medullaris
Ventral complex
THALAMUS

Strionuclear GEP
Posterior striatal NEP and SVZ
Cortical/striatal NEP and SVZ

Putamen (islands)
Internal capsule
Reticular nucleus
Periventricular complex
Thalamic G/EP
Third ventricle
Junction between strionuclear and perifascicular glioepithelia

Temporal NEP and SVZ

TEMPORAL LOBE
Central complex
Perifascicular GEP

Temporal (granular) STF

Internal capsule (posterior limb)
Optic tract
Aqueduct

MIDBRAIN TEGMENTUM
Mesencephalic G/EP

STF1 t2
STF2
STF3a
STF3b
STF3c
STF4 t2
STF5

Caudate nucleus (tail)
Stria terminalis
Cerebral peduncle
Substantia nigra
Interpeduncular nucleus

Lateral ventricle
Dentate gyrus
Ammon's horn
HIPPO-CAMPUS

Middle and inferior temporal gyri
Subgranular zone
Dentate migration
Ammonic migration and sojourn zone

Entorhinal cortex
Parahippocampal gyrus

Hippocampal NEP
Parahippocampal NEP and SVZ

G/EP - glioepithelium/ependyma
GEP - glioepithelium
NEP - neuroepithelium
SVZ - subventricular zone
Germinal and transitional structures in *italics*

PLATE 16A
CR 165 mm, GW 20, Y17-63
Frontal, Section 551

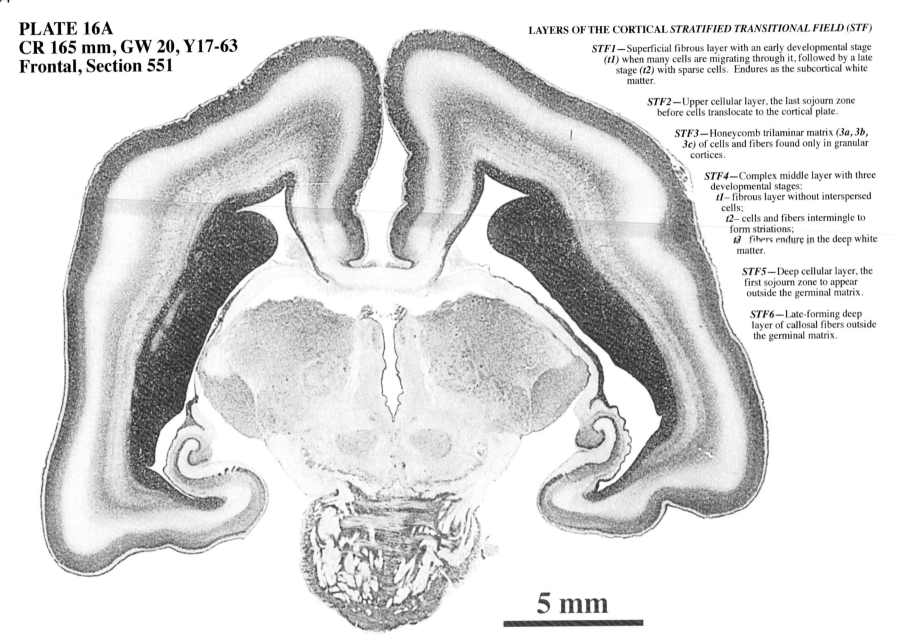

LAYERS OF THE CORTICAL *STRATIFIED TRANSITIONAL FIELD (STF)*

STF1—Superficial fibrous layer with an early developmental stage *(t1)* when many cells are migrating through it, followed by a late stage *(t2)* with sparse cells. Endures as the subcortical white matter.

STF2—Upper cellular layer, the last sojourn zone before cells translocate to the cortical plate.

STF3—Honeycomb trilaminar matrix *(3a, 3b, 3c)* of cells and fibers found only in granular cortices.

STF4—Complex middle layer with three developmental stages:
  t1– fibrous layer without interspersed cells;
  t2– cells and fibers intermingle to form striations;
  t3 fibers endure in the deep white matter.

STF5—Deep cellular layer, the first sojourn zone to appear outside the germinal matrix.

STF6—Late-forming deep layer of callosal fibers outside the germinal matrix.

5 mm

See detail of the brain core in Plates 28A and B.

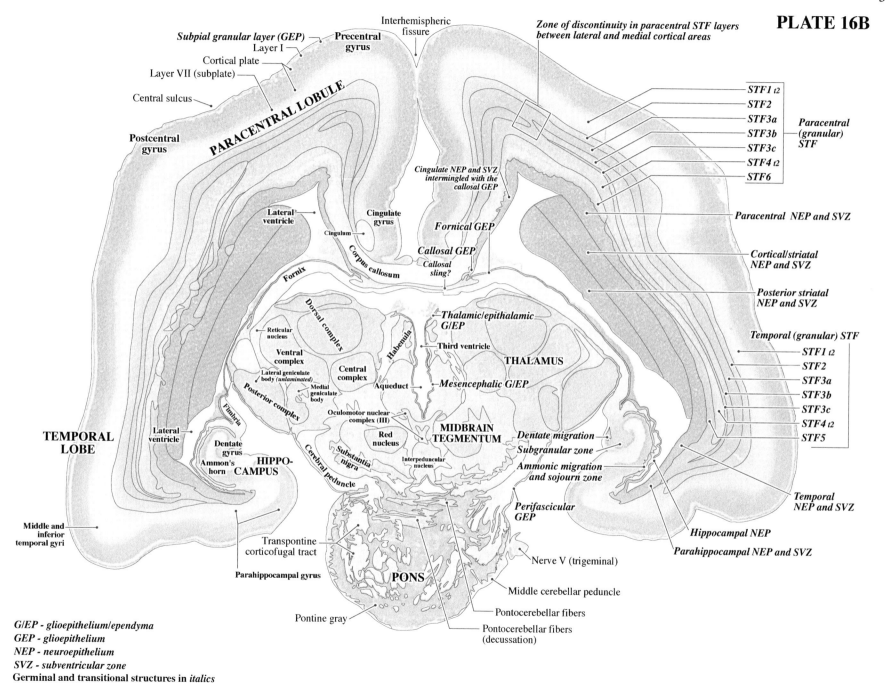

PLATE 16B

35

Interhemispheric fissure

Zone of discontinuity in paracentral STF layers between lateral and medial cortical areas

Subpial granular layer (GEP)
Layer I
Cortical plate
Precentral gyrus
Layer VII (subplate)
Central sulcus

Postcentral gyrus

PARACENTRAL LOBULE

STF1 t2
STF2
STF3a
STF3b
STF3c
STF4 t2
STF6

Paracentral (granular) STF

Cingulate NEP and SVZ intermingled with the callosal GEP

Lateral ventricle
Cingulate gyrus
Cingulum
Corpus callosum
Fornical GEP
Callosal GEP
Callosal sling?

Paracentral NEP and SVZ

Cortical/striatal NEP and SVZ

Posterior striatal NEP and SVZ

Fornix

Dorsal complex
Reticular nucleus
Ventral complex
Lateral geniculate body *(unlaminated)*
Medial geniculate body
Posterior complex

Habenula
Thalamic/epithalamic G/EP
Third ventricle
THALAMUS
Central complex
Aqueduct
Mesencephalic G/EP

Temporal (granular) STF
STF1 t2
STF2
STF3a
STF3b
STF3c
STF4 t2
STF5

TEMPORAL LOBE

Lateral ventricle
Fimbria
Dentate gyrus
Ammon's horn
HIPPO-CAMPUS

Oculomotor nuclear complex (III)
Red nucleus
Substantia nigra
MIDBRAIN TEGMENTUM
Interpeduncular nucleus
Cerebral peduncle

Dentate migration
Subgranular zone
Ammonic migration and sojourn zone
Perifascicular GEP

Temporal NEP and SVZ

Hippocampal NEP
Parahippocampal NEP and SVZ

Middle and inferior temporal gyri

Transpontine corticofugal tract
Parahippocampal gyrus

PONS

Nerve V (trigeminal)
Middle cerebellar peduncle
Pontocerebellar fibers

Pontine gray
Pontocerebellar fibers (decussation)

G/EP - glioepithelium/ependyma
GEP - glioepithelium
NEP - neuroepithelium
SVZ - subventricular zone
Germinal and transitional structures in *italics*

PLATE 17A
CR 165 mm, GW 20, Y17-63
Frontal, Section 601

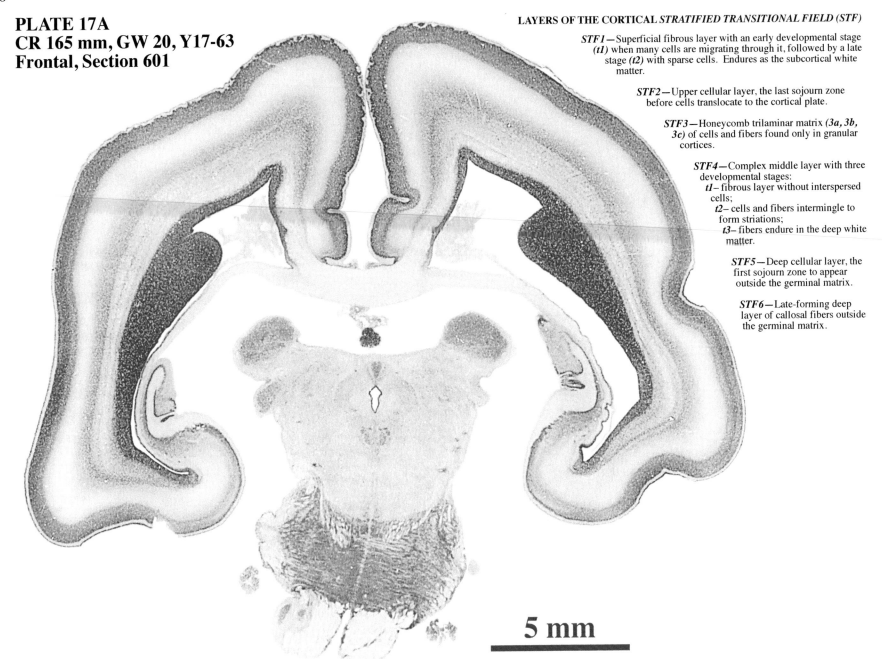

LAYERS OF THE CORTICAL *STRATIFIED TRANSITIONAL FIELD (STF)*

STF1—Superficial fibrous layer with an early developmental stage
(t1) when many cells are migrating through it, followed by a late
stage *(t2)* with sparse cells. Endures as the subcortical white
matter.

STF2—Upper cellular layer, the last sojourn zone
before cells translocate to the cortical plate.

STF3—Honeycomb trilaminar matrix *(3a, 3b,*
3c) of cells and fibers found only in granular
cortices.

STF4—Complex middle layer with three
developmental stages:
 t1– fibrous layer without interspersed
 cells;
  t2– cells and fibers intermingle to
  form striations;
   t3– fibers endure in the deep white
   matter.

STF5—Deep cellular layer, the
first sojourn zone to appear
outside the germinal matrix.

STF6—Late-forming deep
layer of callosal fibers outside
the germinal matrix.

5 mm

See detail of the brain core in Plates 29A and B.

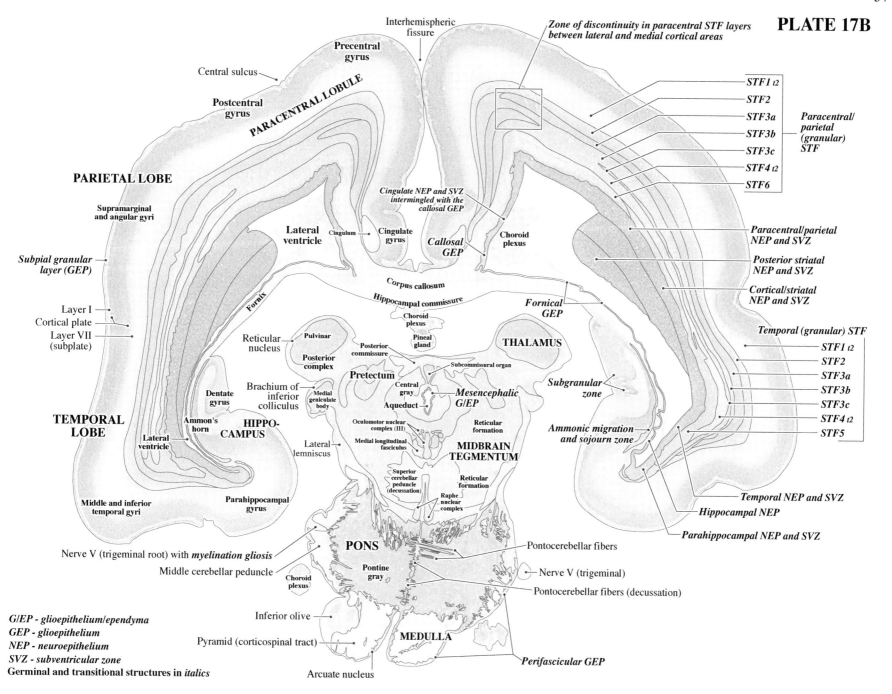

PLATE 17B

Interhemispheric fissure

Zone of discontinuity in paracentral STF layers between lateral and medial cortical areas

Precentral gyrus

Central sulcus

Postcentral gyrus

PARACENTRAL LOBULE

STF1 *t2*
STF2
STF3a
STF3b
STF3c
STF4 *t2*
STF6

Paracentral/ parietal (granular) STF

PARIETAL LOBE

Supramarginal and angular gyri

Cingulate NEP and SVZ intermingled with the callosal GEP

Lateral ventricle

Cingulum

Cingulate gyrus

Callosal GEP

Choroid plexus

Paracentral/parietal NEP and SVZ

Posterior striatal NEP and SVZ

Subpial granular layer (GEP)

Corpus callosum

Hippocampal commissure

Choroid plexus

Fornical GEP

Cortical/striatal NEP and SVZ

Layer I
Cortical plate
Layer VII (subplate)

Fornix

Reticular nucleus

Pulvinar

Posterior commissure

Pineal gland

THALAMUS

Temporal (granular) STF

Posterior complex

Subcommissural organ

Subgranular zone

STF1 *t2*
STF2
STF3a
STF3b
STF3c
STF4 *t2*
STF5

Brachium of inferior colliculus

Dentate gyrus

Medial geniculate body

Pretectum

Central gray

Aqueduct

Mesencephalic G/EP

TEMPORAL LOBE

Ammon's horn

HIPPO-CAMPUS

Oculomotor nuclear complex (III)

Reticular formation

MIDBRAIN TEGMENTUM

Ammonic migration and sojourn zone

Lateral ventricle

Lateral lemniscus

Medial longitudinal fasciculus

Superior cerebellar peduncle (decussation)

Reticular formation

Middle and inferior temporal gyri

Parahippocampal gyrus

Raphe nuclear complex

Temporal NEP and SVZ

Hippocampal NEP

Pontocerebellar fibers

Parahippocampal NEP and SVZ

Nerve V (trigeminal root) with *myelination gliosis*

PONS

Nerve V (trigeminal)

Middle cerebellar peduncle

Choroid plexus

Pontine gray

Pontocerebellar fibers (decussation)

G/EP - *glioepithelium/ependyma*
GEP - *glioepithelium*
NEP - *neuroepithelium*
SVZ - *subventricular zone*
Germinal and transitional structures in *italics*

Inferior olive

MEDULLA

Pyramid (corticospinal tract)

Perifascicular GEP

Arcuate nucleus

38

PLATE 18A
CR 165 mm, GW 20, Y17-63
Frontal, Section 631

LAYERS OF THE CORTICAL *STRATIFIED TRANSITIONAL FIELD (STF)*

STF1—Superficial fibrous layer with an early developmental stage *(t1)* when many cells are migrating through it, followed by a late stage *(t2)* with sparse cells. Endures as the subcortical white matter.

STF2—Upper cellular layer, the last sojourn zone before cells translocate to the cortical plate.

STF3—Honeycomb trilaminar matrix *(3a, 3b, 3c)* of cells and fibers found only in granular cortices.

STF4—Complex middle layer with three developmental stages:
  t1– fibrous layer without interspersed cells;
   t2– cells and fibers intermingle to form striations;
    t3– fibers endure in the deep white matter.

STF5—Deep cellular layer, the first sojourn zone to appear outside the germinal matrix.

STF6—Late-forming deep layer of callosal fibers outside the germinal matrix.

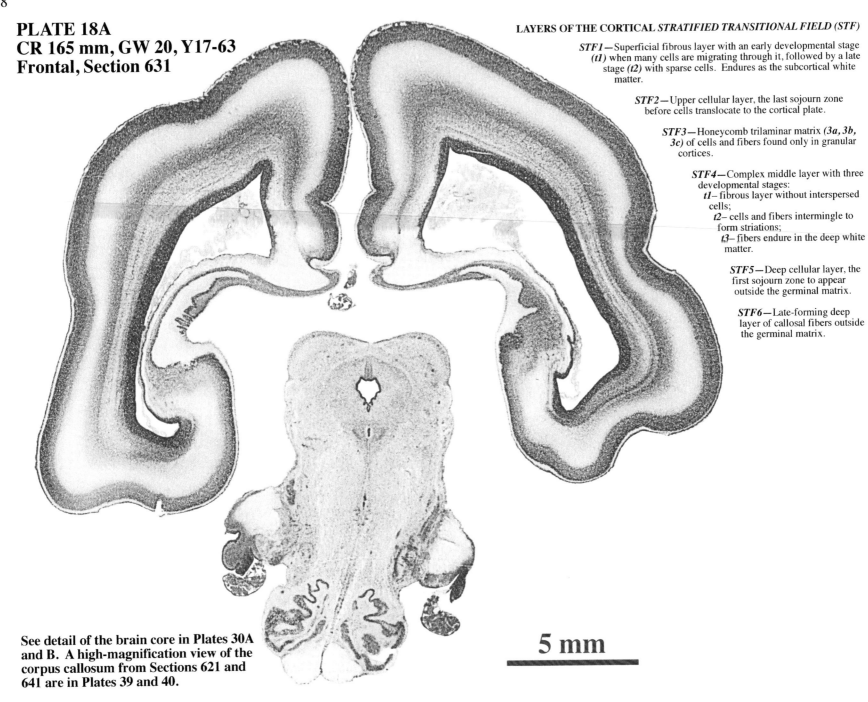

See detail of the brain core in Plates 30A and B. A high-magnification view of the corpus callosum from Sections 621 and 641 are in Plates 39 and 40.

5 mm

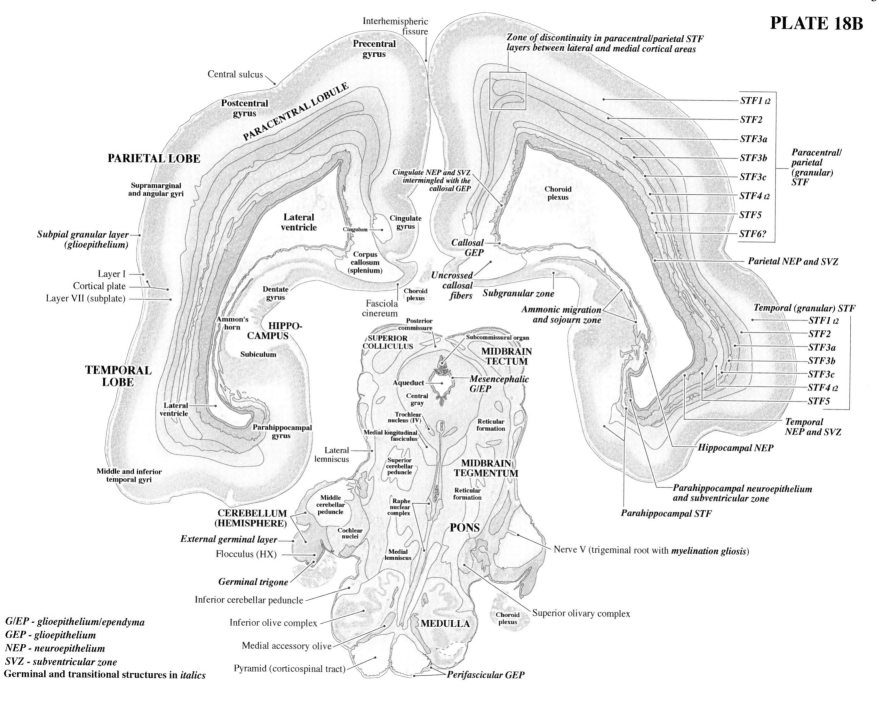

Interhemispheric fissure

Precentral gyrus

Zone of discontinuity in paracentral/parietal STF layers between lateral and medial cortical areas

Central sulcus

Postcentral gyrus

PARACENTRAL LOBULE

PARIETAL LOBE

Supramarginal and angular gyri

Cingulate NEP and SVZ intermingled with the callosal GEP

Choroid plexus

STF1 t2
STF2
STF3a
STF3b
STF3c
STF4 t2
STF5
STF6?

Paracentral/ parietal (granular) STF

Lateral ventricle

Cingulate gyrus

Cingulum

Subpial granular layer (glioepithelium)

Corpus callosum (splenium)

Callosal GEP

Layer I
Cortical plate
Layer VII (subplate)

Dentate gyrus

Choroid plexus

Uncrossed callosal fibers

Subgranular zone

Parietal NEP and SVZ

Fasciola cinereum

HIPPO-CAMPUS

Ammon's horn

Subiculum

Posterior commissure

Subcommissural organ

SUPERIOR COLLICULUS

MIDBRAIN TECTUM

Ammonic migration and sojourn zone

Temporal (granular) STF

STF1 t2
STF2
STF3a
STF3b
STF3c
STF4 t2
STF5

TEMPORAL LOBE

Aqueduct

Central gray

Mesencephalic G/EP

Lateral ventricle

Parahippocampal gyrus

Trochlear nucleus (IV)

Medial longitudinal fasciculus

Reticular formation

Temporal NEP and SVZ

Lateral lemniscus

Superior cerebellar peduncle

MIDBRAIN TEGMENTUM

Hippocampal NEP

Middle and inferior temporal gyri

Middle cerebellar peduncle

Raphe nuclear complex

Reticular formation

Parahippocampal neuroepithelium and subventricular zone

CEREBELLUM (HEMISPHERE)

Cochlear nuclei

PONS

Parahippocampal STF

External germinal layer

Flocculus (HX)

Medial lemniscus

Nerve V (trigeminal root with *myelination gliosis*)

Germinal trigone

Inferior cerebellar peduncle

Inferior olive complex

Choroid plexus

Superior olivary complex

MEDULLA

Medial accessory olive

Pyramid (corticospinal tract)

Perifascicular GEP

G/EP - glioepithelium/ependyma
GEP - glioepithelium
NEP - neuroepithelium
SVZ - subventricular zone
Germinal and transitional structures in *italics*

PLATE 19A
CR 165 mm, GW 20, Y17-63
Frontal, Section 661

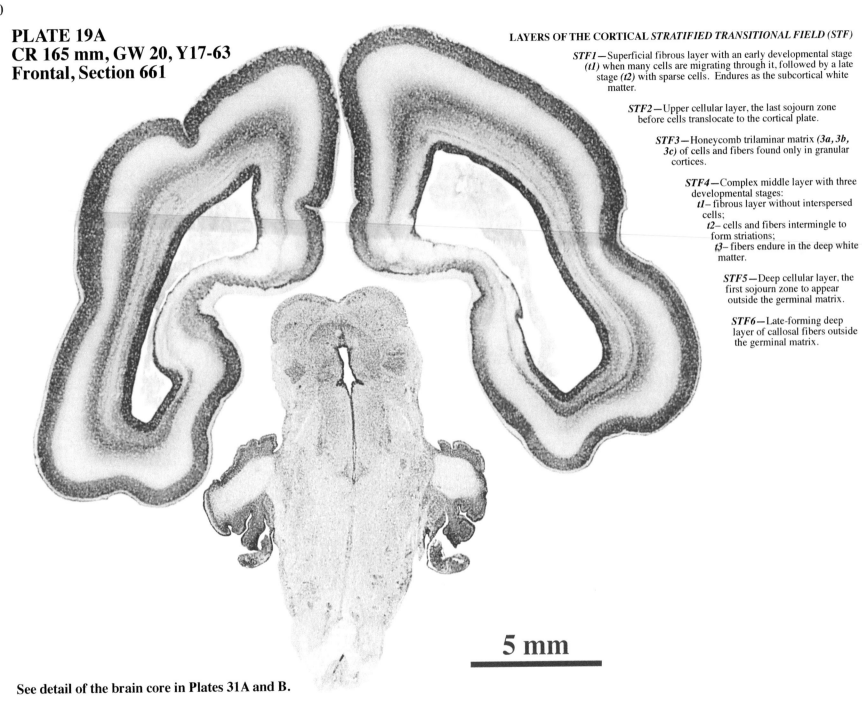

LAYERS OF THE CORTICAL *STRATIFIED TRANSITIONAL FIELD (STF)*

STF1—Superficial fibrous layer with an early developmental stage *(t1)* when many cells are migrating through it, followed by a late stage *(t2)* with sparse cells. Endures as the subcortical white matter.

STF2—Upper cellular layer, the last sojourn zone before cells translocate to the cortical plate.

STF3—Honeycomb trilaminar matrix *(3a, 3b, 3c)* of cells and fibers found only in granular cortices.

STF4—Complex middle layer with three developmental stages:
  t1– fibrous layer without interspersed cells;
  t2– cells and fibers intermingle to form striations;
  t3– fibers endure in the deep white matter.

STF5—Deep cellular layer, the first sojourn zone to appear outside the germinal matrix.

STF6—Late-forming deep layer of callosal fibers outside the germinal matrix.

5 mm

See detail of the brain core in Plates 31A and B.

41

PLATE 19B

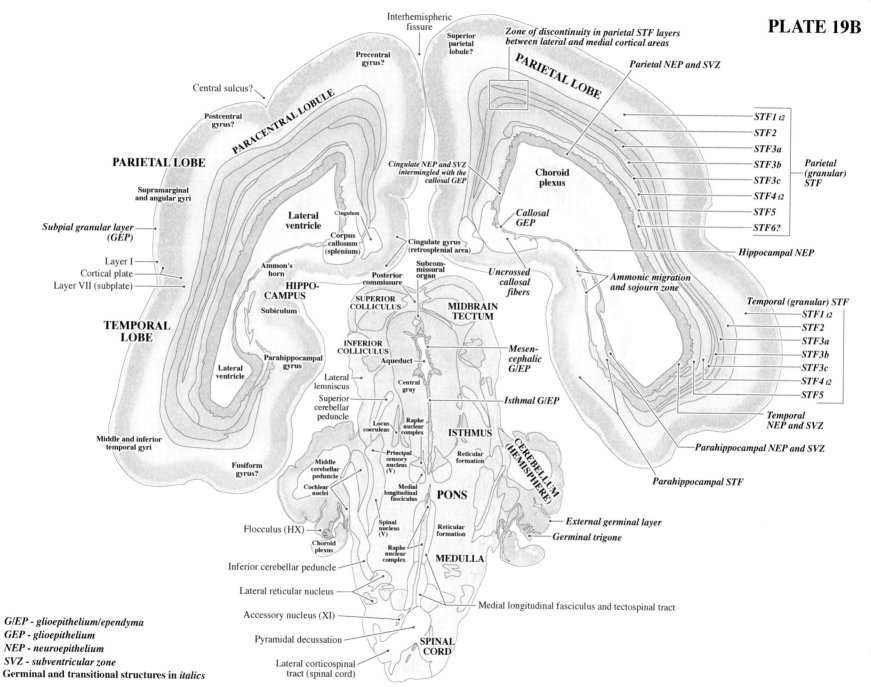

Interhemispheric fissure

Zone of discontinuity in parietal STF layers between lateral and medial cortical areas

Superior parietal lobule?

Precentral gyrus?

PARIETAL LOBE

Parietal NEP and SVZ

Central sulcus?

PARACENTRAL LOBULE

Postcentral gyrus?

PARIETAL LOBE

STF1 t2
STF2
STF3a
STF3b
STF3c
STF4 t2
STF5
STF6?

Parietal (granular) STF

Cingulate NEP and SVZ intermingled with the callosal GEP

Choroid plexus

Supramarginal and angular gyri

Lateral ventricle

Cingulum

Callosal GEP

Subpial granular layer (GEP)

Corpus callosum (splenium)

Cingulate gyrus (retrosplenial area)

Hippocampal NEP

Layer I

Ammon's horn

HIPPO-CAMPUS

Posterior commissure

Subcommissural organ

Uncrossed callosal fibers

Ammonic migration and sojourn zone

Cortical plate

Layer VII (subplate)

TEMPORAL LOBE

Subiculum

SUPERIOR COLLICULUS

MIDBRAIN TECTUM

Temporal (granular) STF

STF1 t2
STF2
STF3a
STF3b
STF3c
STF4 t2
STF5

Parahippocampal gyrus

INFERIOR COLLICULUS

Aqueduct

Mesencephalic G/EP

Lateral ventricle

Lateral lemniscus

Central gray

Isthmal G/EP

Temporal NEP and SVZ

Middle and inferior temporal gyri

Superior cerebellar peduncle

Locus coeruleus

Raphe nuclear complex

ISTHMUS

CEREBELLUM (HEMISPHERE)

Parahippocampal NEP and SVZ

Fusiform gyrus?

Middle cerebellar peduncle

Cochlear nuclei

Principal sensory nucleus (V)

Reticular formation

Parahippocampal STF

Medial longitudinal fasciculus

PONS

Flocculus (HX)

Spinal nucleus (V)

Reticular formation

External germinal layer

Choroid plexus

Raphe nuclear complex

MEDULLA

Germinal trigone

Inferior cerebellar peduncle

Lateral reticular nucleus

Medial longitudinal fasciculus and tectospinal tract

Accessory nucleus (XI)

Pyramidal decussation

SPINAL CORD

Lateral corticospinal tract (spinal cord)

G/EP - glioepithelium/ependyma
GEP - glioepithelium
NEP - neuroepithelium
SVZ - subventricular zone
Germinal and transitional structures in *italics*

42

PLATE 20A
CR 165 mm, GW 20, Y17-63
Frontal, Section 701

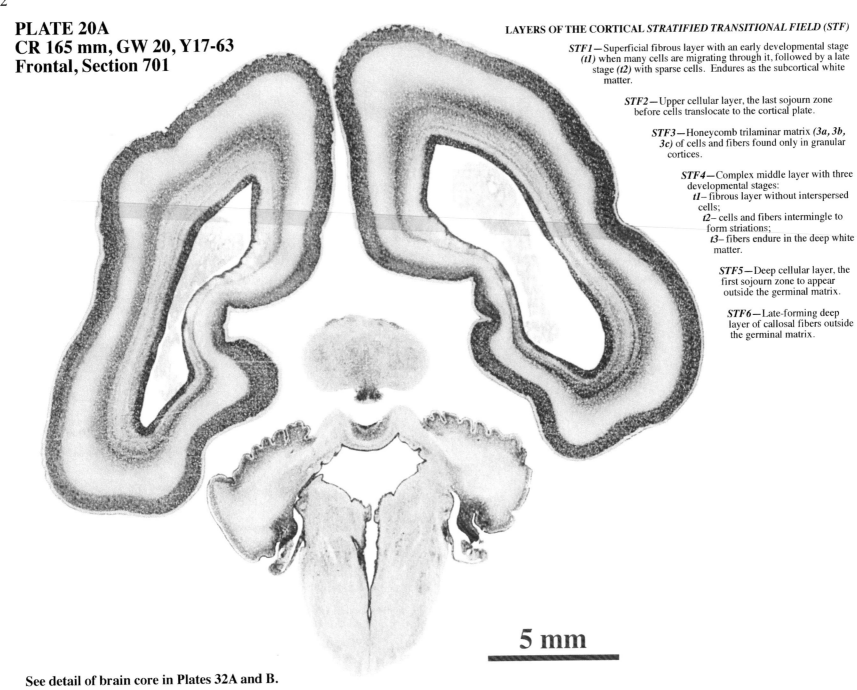

LAYERS OF THE CORTICAL *STRATIFIED TRANSITIONAL FIELD (STF)*

STF1—Superficial fibrous layer with an early developmental stage *(t1)* when many cells are migrating through it, followed by a late stage *(t2)* with sparse cells. Endures as the subcortical white matter.

STF2—Upper cellular layer, the last sojourn zone before cells translocate to the cortical plate.

STF3—Honeycomb trilaminar matrix *(3a, 3b, 3c)* of cells and fibers found only in granular cortices.

STF4—Complex middle layer with three developmental stages:
  t1–fibrous layer without interspersed cells;
  t2–cells and fibers intermingle to form striations;
  t3–fibers endure in the deep white matter.

STF5—Deep cellular layer, the first sojourn zone to appear outside the germinal matrix.

STF6—Late-forming deep layer of callosal fibers outside the germinal matrix.

5 mm

See detail of brain core in Plates 32A and B.

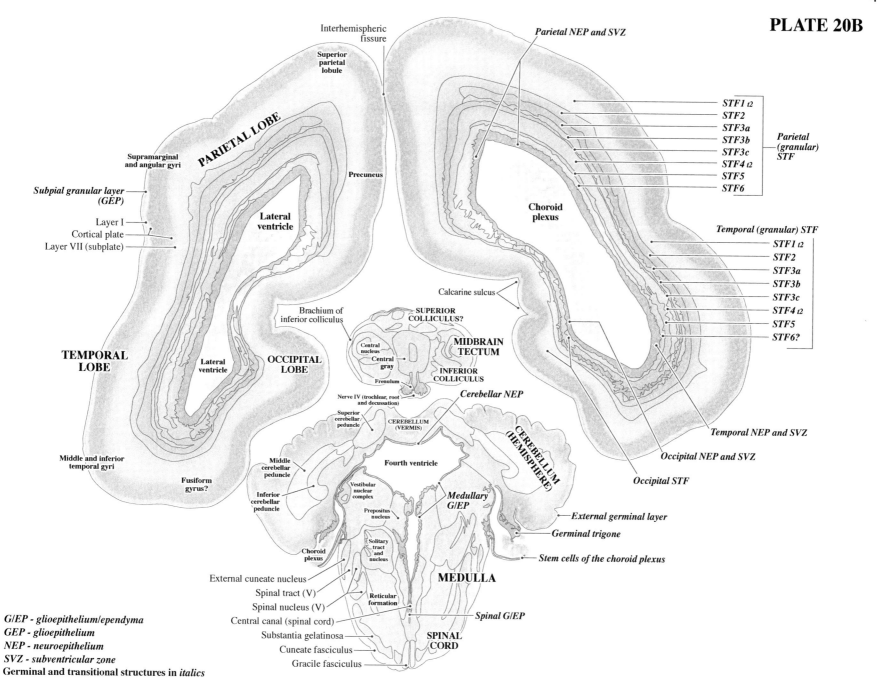

Interhemispheric fissure

Superior parietal lobule

Parietal NEP and SVZ

PARIETAL LOBE

Supramarginal and angular gyri

Precuneus

Subpial granular layer (GEP)

Lateral ventricle

Layer I
Cortical plate
Layer VII (subplate)

Choroid plexus

STF1 t2
STF2
STF3a
STF3b
STF3c
STF4 t2
STF5
STF6

Parietal (granular) STF

Temporal (granular) STF

STF1 t2
STF2
STF3a
STF3b
STF3c
STF4 t2
STF5
STF6?

TEMPORAL LOBE

Lateral ventricle

OCCIPITAL LOBE

Brachium of inferior colliculus

SUPERIOR COLLICULUS?

Central nucleus
Central gray

MIDBRAIN TECTUM

INFERIOR COLLICULUS

Frenulum

Calcarine sulcus

Middle and inferior temporal gyri

Fusiform gyrus?

Nerve IV (trochlear, root and decussation)

Superior cerebellar peduncle

Cerebellar NEP

CEREBELLUM (VERMIS)

CEREBELLUM (HEMISPHERE)

Middle cerebellar peduncle

Fourth ventricle

Vestibular nuclear complex

Prepositus nucleus

Medullary G/EP

Inferior cerebellar peduncle

Choroid plexus

Solitary tract and nucleus

External cuneate nucleus

Spinal tract (V)

Spinal nucleus (V)

Reticular formation

Central canal (spinal cord)

Spinal G/EP

MEDULLA

Temporal NEP and SVZ

Occipital NEP and SVZ

Occipital STF

External germinal layer
Germinal trigone

Stem cells of the choroid plexus

Substantia gelatinosa
Cuneate fasciculus
Gracile fasciculus

SPINAL CORD

G/EP - glioepithelium/ependyma
GEP - glioepithelium
NEP - neuroepithelium
SVZ - subventricular zone
Germinal and transitional structures in *italics*

PLATE 21A
CR 165 mm, GW 20, Y17-63
Frontal, Section 741

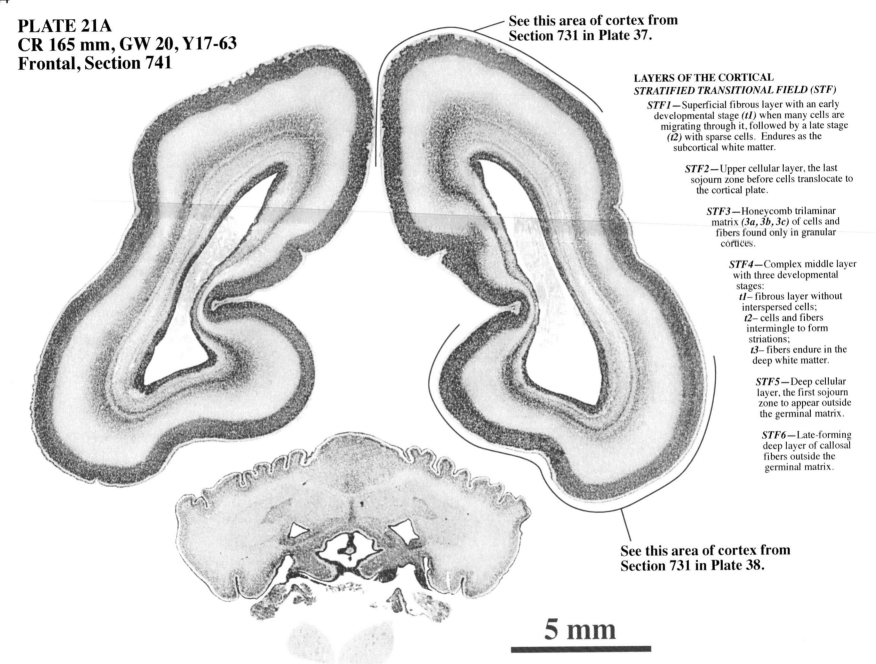

See this area of cortex from
Section 731 in Plate 37.

LAYERS OF THE CORTICAL
STRATIFIED TRANSITIONAL FIELD (STF)

STF1—Superficial fibrous layer with an early
developmental stage *(t1)* when many cells are
migrating through it, followed by a late stage
(t2) with sparse cells. Endures as the
subcortical white matter.

STF2—Upper cellular layer, the last
sojourn zone before cells translocate to
the cortical plate.

STF3—Honeycomb trilaminar
matrix *(3a, 3b, 3c)* of cells and
fibers found only in granular
cortices.

STF4—Complex middle layer
with three developmental
stages:
  t1– fibrous layer without
  interspersed cells;
  t2– cells and fibers
  intermingle to form
  striations;
  t3– fibers endure in the
  deep white matter.

STF5—Deep cellular
layer, the first sojourn
zone to appear outside
the germinal matrix.

STF6—Late-forming
deep layer of callosal
fibers outside the
germinal matrix.

See this area of cortex from
Section 731 in Plate 38.

5 mm

See detail of the brain core in Plates 33A and B.

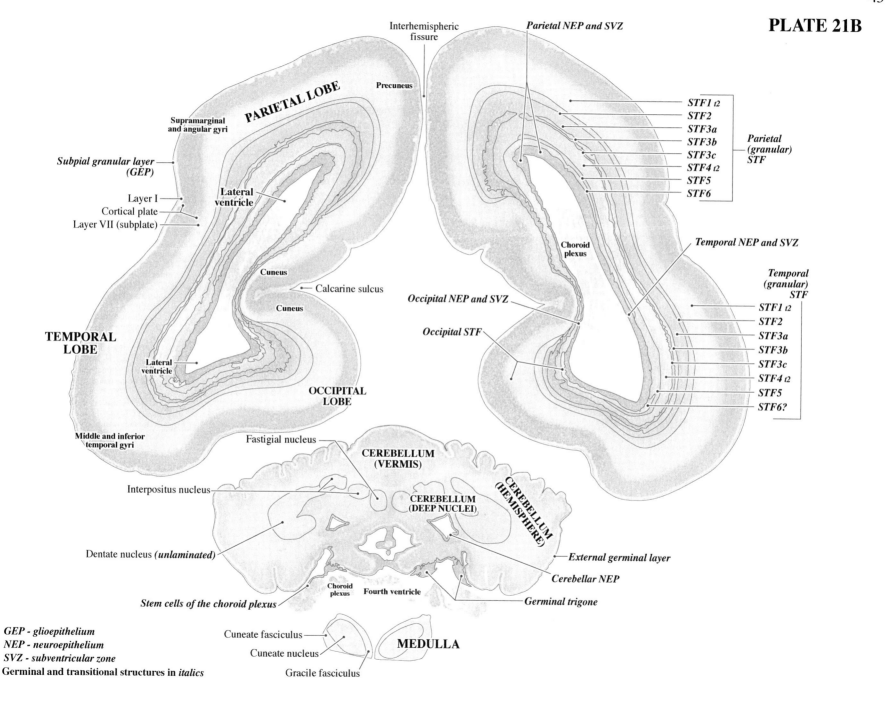

Interhemispheric fissure

Parietal NEP and SVZ

Precuneus

PARIETAL LOBE

Supramarginal and angular gyri

STF1 t2
STF2
STF3a
STF3b
STF3c
STF4 t2
STF5
STF6

Parietal (granular) STF

Subpial granular layer (GEP)

Layer I
Cortical plate
Layer VII (subplate)

Lateral ventricle

Choroid plexus

Temporal NEP and SVZ

TEMPORAL LOBE

Cuneus

Calcarine sulcus

Cuneus

Occipital NEP and SVZ

Occipital STF

Temporal (granular) STF

STF1 t2
STF2
STF3a
STF3b
STF3c
STF4 t2
STF5
STF6?

Lateral ventricle

OCCIPITAL LOBE

Middle and inferior temporal gyri

Fastigial nucleus

CEREBELLUM (VERMIS)

CEREBELLUM (HEMISPHERE)

Interpositus nucleus

CEREBELLUM (DEEP NUCLEI)

Dentate nucleus *(unlaminated)*

External germinal layer

Cerebellar NEP

Stem cells of the choroid plexus

Choroid plexus

Fourth ventricle

Germinal trigone

Cuneate fasciculus

MEDULLA

Cuneate nucleus

Gracile fasciculus

GEP - glioepithelium
NEP - neuroepithelium
SVZ - subventricular zone
Germinal and transitional structures in *italics*

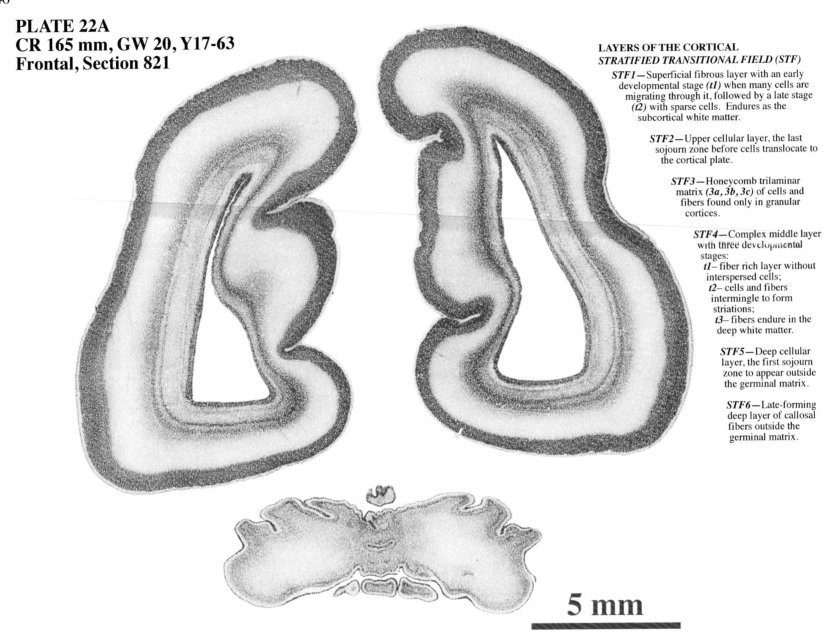

PLATE 22A
CR 165 mm, GW 20, Y17-63
Frontal, Section 821

LAYERS OF THE CORTICAL
STRATIFIED TRANSITIONAL FIELD (STF)

STF1—Superficial fibrous layer with an early developmental stage *(t1)* when many cells are migrating through it, followed by a late stage *(t2)* with sparse cells. Endures as the subcortical white matter.

STF2—Upper cellular layer, the last sojourn zone before cells translocate to the cortical plate.

STF3—Honeycomb trilaminar matrix *(3a, 3b, 3c)* of cells and fibers found only in granular cortices.

STF4—Complex middle layer with three developmental stages:
  t1– fiber rich layer without interspersed cells;
  t2– cells and fibers intermingle to form striations;
  t3– fibers endure in the deep white matter.

STF5—Deep cellular layer, the first sojourn zone to appear outside the germinal matrix.

STF6—Late-forming deep layer of callosal fibers outside the germinal matrix.

5 mm

See detail of the brain core in Plates 34A and B.

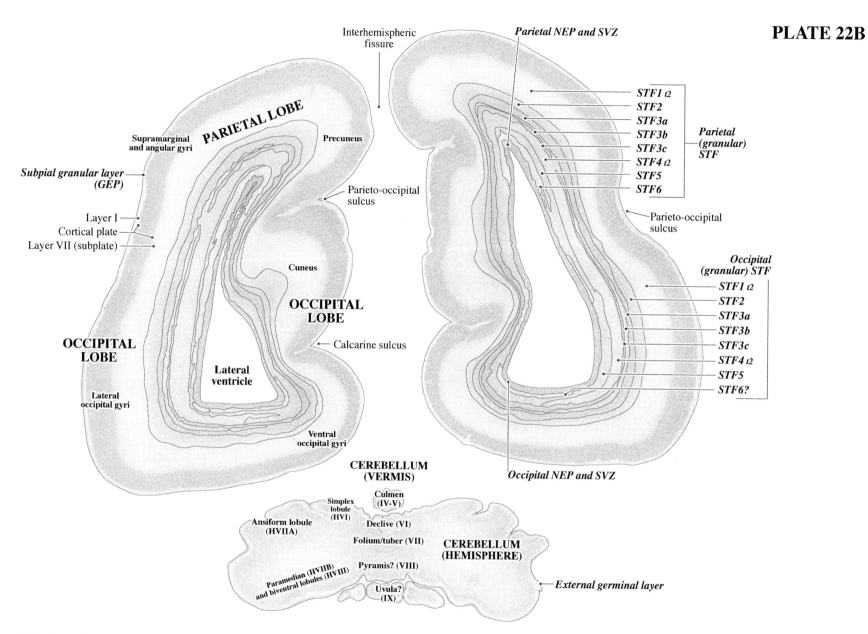

Interhemispheric fissure

Parietal NEP and SVZ

PARIETAL LOBE

Supramarginal and angular gyri

Precuneus

Subpial granular layer (GEP)

Parieto-occipital sulcus

Layer I
Cortical plate
Layer VII (subplate)

Cuneus

OCCIPITAL LOBE

OCCIPITAL LOBE

Lateral occipital gyri

Calcarine sulcus

Lateral ventricle

Ventral occipital gyri

STF1 t2
STF2
STF3a
STF3b
STF3c
STF4 t2
STF5
STF6

Parietal (granular) STF

Parieto-occipital sulcus

Occipital (granular) STF

STF1 t2
STF2
STF3a
STF3b
STF3c
STF4 t2
STF5
STF6?

Occipital NEP and SVZ

CEREBELLUM (VERMIS)

Culmen (IV-V)
Simplex lobule (HVI)
Ansiform lobule (HVIIA)
Declive (VI)
Folium/tuber (VII)
CEREBELLUM (HEMISPHERE)
Paramedian (HVIIB) and biventral lobules (HVIII)
Pyramis? (VIII)
Uvula? (IX)
External germinal layer

GEP - glioepithelium
NEP - neuroepithelium
SVZ - subventricular zone
Germinal and transitional structures in *italics*

48

CR 165 mm
GW 20, Y17-63
Frontal
Section 281

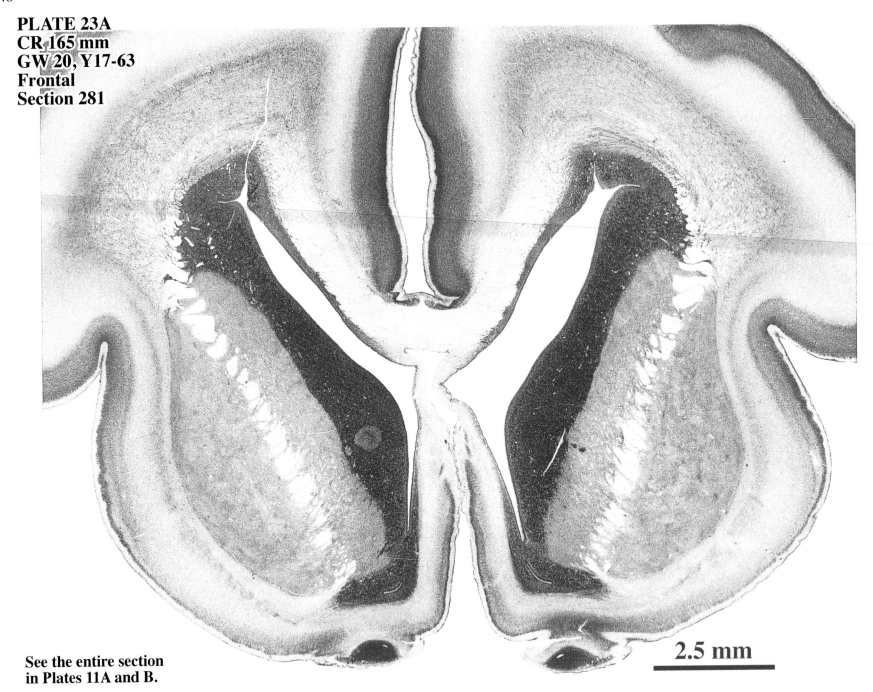

2.5 mm

See the entire section
in Plates 11A and B.

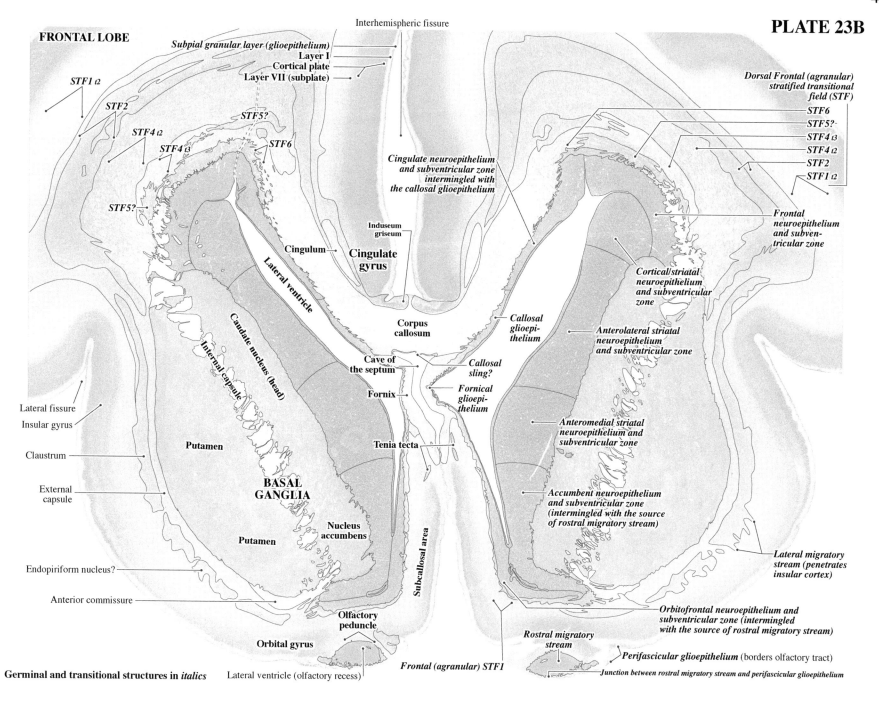

FRONTAL LOBE

Interhemispheric fissure

Subpial granular layer (glioepithelium)
Layer I
Cortical plate
Layer VII (subplate)

STF1 t2

STF2

STF4 t2

STF4 t3

STF5?

STF6

STF5?

Dorsal Frontal (agranular) stratified transitional field (STF)

STF6
STF5?
STF4 t3
STF4 t2
STF2
STF1 t2

Frontal neuroepithelium and subventricular zone

Cingulate neuroepithelium and subventricular zone intermingled with the callosal glioepithelium

Induseum griseum

Cingulum

Cingulate gyrus

Lateral ventricle

Caudate nucleus (head)

Internal capsule

Corpus callosum

Callosal glioepithelium

Cortical/striatal neuroepithelium and subventricular zone

Anterolateral striatal neuroepithelium and subventricular zone

Cave of the septum

Callosal sling?

Fornix

Fornical glioepithelium

Anteromedial striatal neuroepithelium and subventricular zone

Lateral fissure

Insular gyrus

Claustrum

External capsule

Putamen

BASAL GANGLIA

Putamen

Tenia tecta

Nucleus accumbens

Subcallosal area

Accumbent neuroepithelium and subventricular zone (intermingled with the source of rostral migratory stream)

Lateral migratory stream (penetrates insular cortex)

Endopiriform nucleus?

Anterior commissure

Olfactory peduncle

Orbital gyrus

Orbitofrontal neuroepithelium and subventricular zone (intermingled with the source of rostral migratory stream)

Rostral migratory stream

Perifascicular glioepithelium (borders olfactory tract)

Germinal and transitional structures in *italics*

Lateral ventricle (olfactory recess)

Frontal (agranular) STF1

Junction between rostral migratory stream and perifascicular glioepithelium

PLATE 24A
CR 165 mm
GW 20, Y17-63
Frontal
Section 361

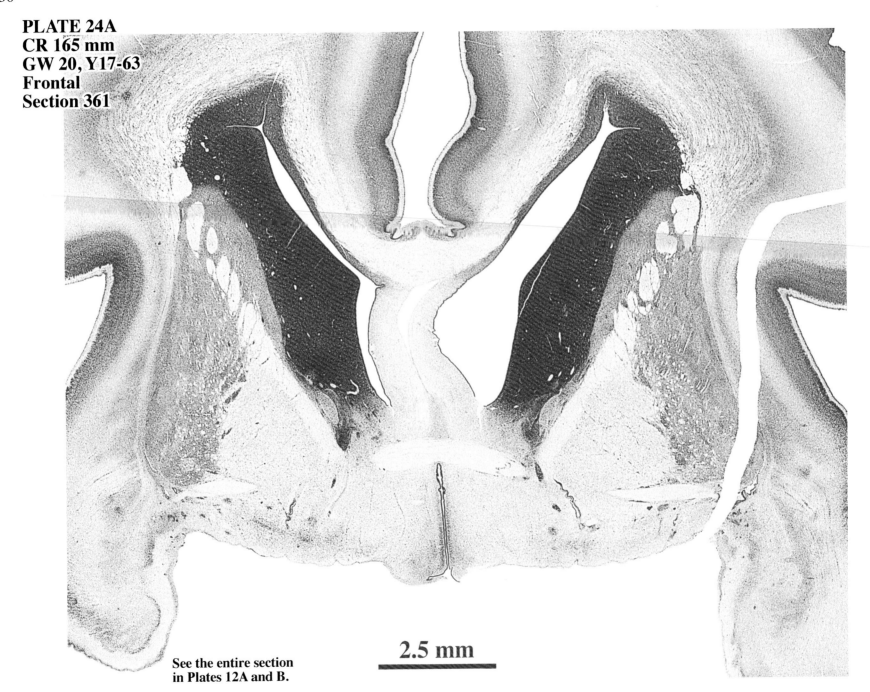

2.5 mm

See the entire section
in Plates 12A and B.

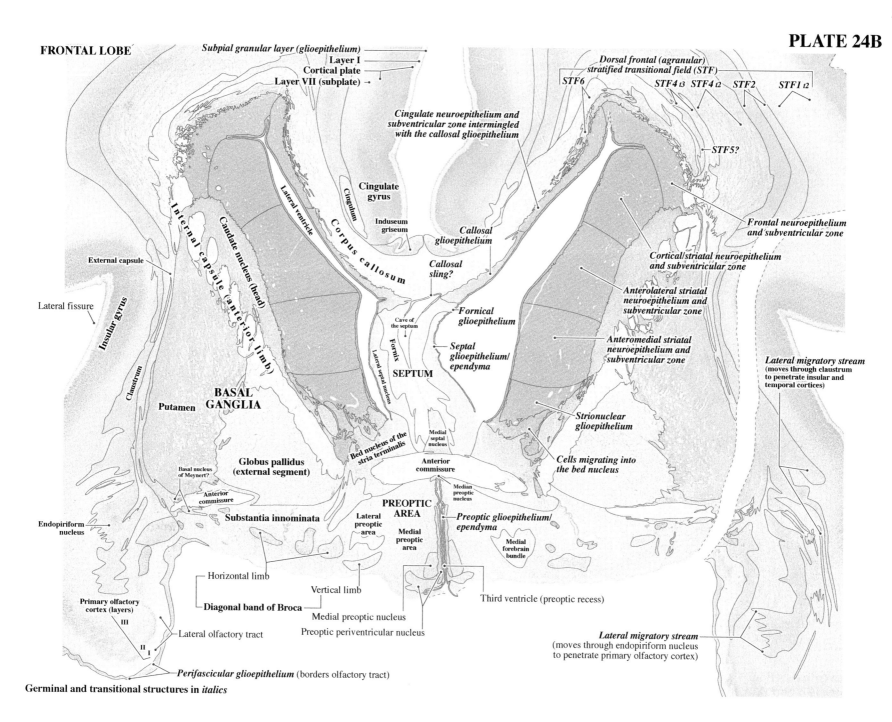

FRONTAL LOBE

Subpial granular layer (glioepithelium)

Layer I

Cortical plate

Layer VII (subplate) →

Dorsal frontal (agranular) stratified transitional field (STF)

STF6

STF4 t3   *STF4 t2*   *STF2*   *STF1 t2*

STF5?

Cingulate neuroepithelium and subventricular zone intermingled with the callosal glioepithelium

Cingulate gyrus

Cingulum

Induseum griseum

Callosal glioepithelium

Callosal sling?

Lateral ventricle

Corpus callosum

Frontal neuroepithelium and subventricular zone

Cortical/striatal neuroepithelium and subventricular zone

Anterolateral striatal neuroepithelium and subventricular zone

I n t e r n a l   c a p s u l e

Caudate nucleus (head)

External capsule

Fornical glioepithelium

Anteromedial striatal neuroepithelium and subventricular zone

Lateral fissure

Insular gyrus

Internal capsule (anterior limb)

Cave of the septum

Septal glioepithelium/ ependyma

Lateral migratory stream (moves through claustrum to penetrate insular and temporal cortices)

Claustrum

Fornix

Lateral septal nucleus

SEPTUM

Strionuclear glioepithelium

BASAL GANGLIA

Putamen

Basal nucleus of Meynert?

Globus pallidus (external segment)

Bed nucleus of the stria terminalis

Medial septal nucleus

Anterior commissure

Cells migrating into the bed nucleus

Anterior commissure

Median preoptic nucleus

Endopiriform nucleus

Substantia innominata

Lateral preoptic area

PREOPTIC AREA

Medial preoptic area

Preoptic glioepithelium/ ependyma

Lateral migratory stream (moves through endopiriform nucleus to penetrate primary olfactory cortex)

Medial forebrain bundle

Horizontal limb

Vertical limb

Third ventricle (preoptic recess)

Primary olfactory cortex (layers)

III

Diagonal band of Broca

Medial preoptic nucleus

II
1

Lateral olfactory tract

Preoptic periventricular nucleus

Perifascicular glioepithelium (borders olfactory tract)

Germinal and transitional structures in *italics*

PLATE 25A
CR 165 mm
GW 20, Y17-63
Frontal
Section 441

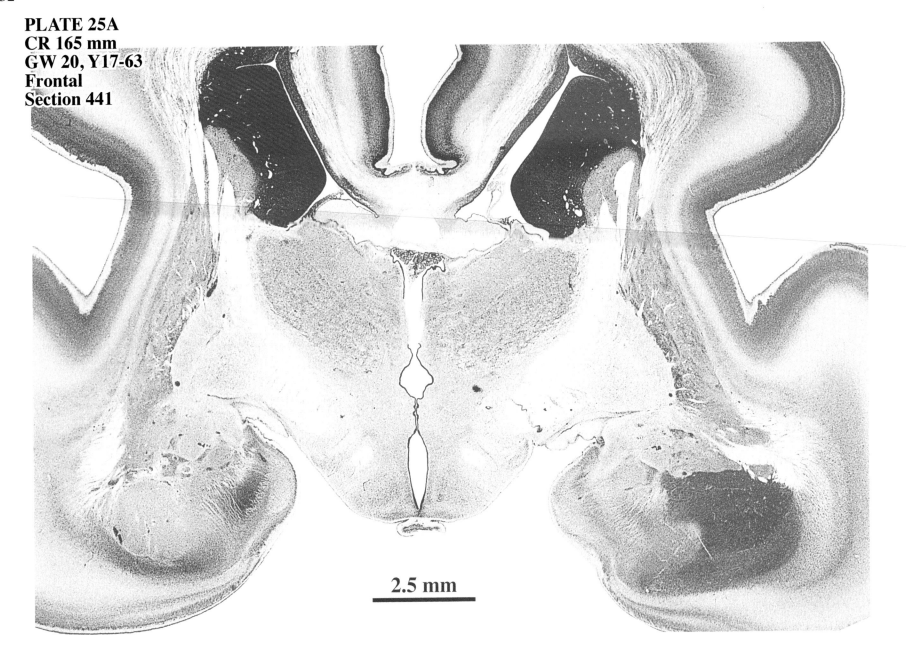

2.5 mm

See the entire section
in Plates 13A and B.

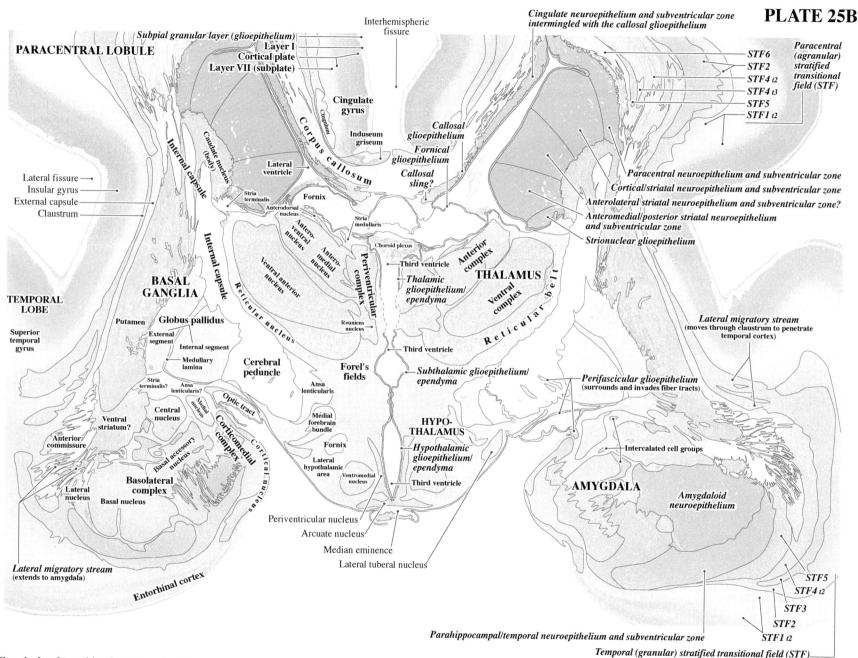

PARACENTRAL LOBULE

Interhemispheric fissure

Cingulate neuroepithelium and subventricular zone intermingled with the callosal glioepithelium

Subpial granular layer (glioepithelium)
Layer I
Cortical plate
Layer VII (subplate)

STF6
STF2
STF4 t2
STF4 t3
STF5
STF1 t2

Paracentral (agranular) stratified transitional field (STF)

Cingulate gyrus

Cingulum

Corpus callosum

Induseum griseum

Callosal glioepithelium

Fornical glioepithelium

Callosal sling?

Caudate nucleus (body)

Internal capsule

Lateral fissure
Insular gyrus
External capsule
Claustrum

Lateral ventricle

Stria terminalis

Fornix

Anterodorsal nucleus

Antero-ventral nucleus

Stria medullaris

Paracentral neuroepithelium and subventricular zone

Cortical/striatal neuroepithelium and subventricular zone

Anterolateral striatal neuroepithelium and subventricular zone?

Anteromedial/posterior striatal neuroepithelium and subventricular zone

Strionuclear glioepithelium

Internal capsule

BASAL GANGLIA

Antero-medial nucleus

Ventral anterior nucleus

Reticular nucleus

Choroid plexus

Periventricular complex

Third ventricle

Anterior complex

THALAMUS

Thalamic glioepithelium/ ependyma

Ventral complex

Reticular belt

TEMPORAL LOBE

Superior temporal gyrus

Putamen
Globus pallidus

External segment

Internal segment

Medullary lamina

Reuniens nucleus

Third ventricle

Lateral migratory stream (moves through claustrum to penetrate temporal cortex)

Cerebral peduncle

Forel's fields

Subthalamic glioepithelium/ ependyma

Perifascicular glioepithelium (surrounds and invades fiber tracts)

Stria terminalis?

Ansa lenticularis?

Optic tract

Ansa lenticularis

Central nucleus

Medial nucleus

Corticomedial complex

Ventral striatum?

Anterior commissure

Basal accessory nucleus

Medial forebrain bundle

Fornix

Lateral hypothalamic area

HYPO-THALAMUS

Cortical nuclei

Ventromedial nucleus

Hypothalamic glioepithelium/ ependyma

Third ventricle

Intercalated cell groups

AMYGDALA

Amygdaloid neuroepithelium

Basolateral complex

Lateral nucleus

Basal nucleus

Periventricular nucleus

Arcuate nucleus

Median eminence

Lateral tuberal nucleus

Lateral migratory stream (extends to amygdala)

Entorhinal cortex

STF5
STF4 t2
STF3
STF2
STF1 t2

Parahippocampal/temporal neuroepithelium and subventricular zone

Temporal (granular) stratified transitional field (STF)

Germinal and transitional structures in *italics*

PLATE 26A
CR 165 mm
GW 20, Y17-63
Frontal
Section 481

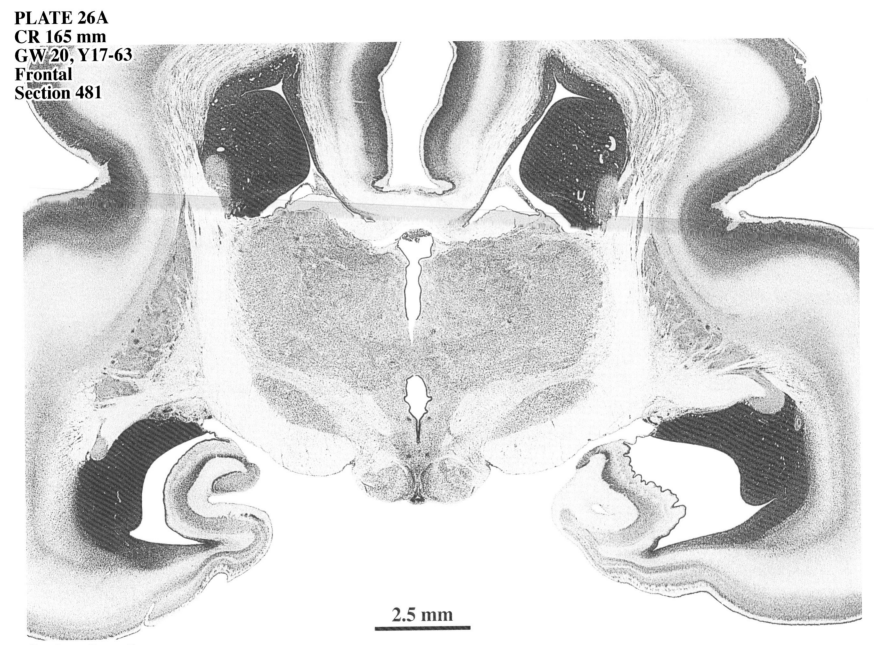

2.5 mm

**See the entire section
in Plates 14A and B.**

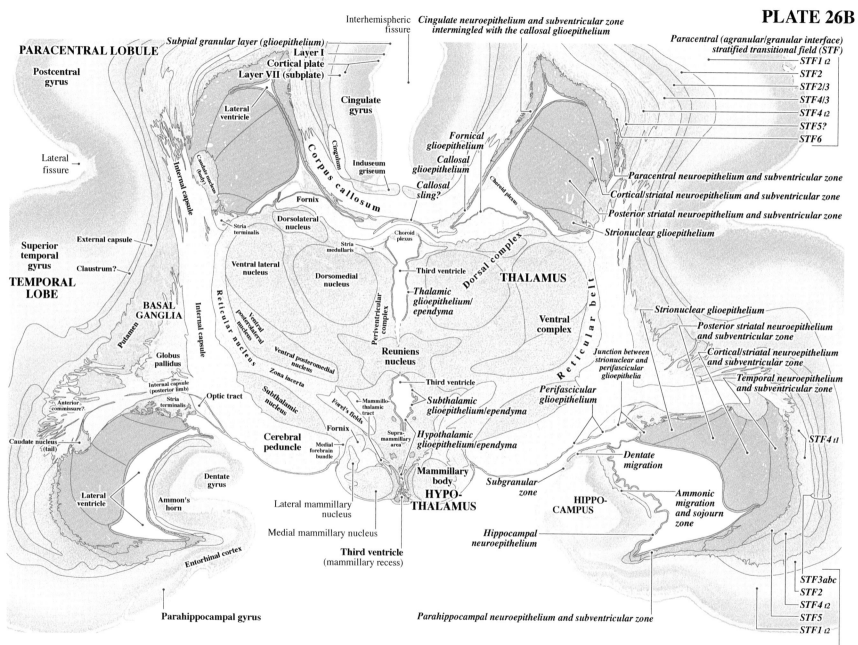

Interhemispheric fissure

Cingulate neuroepithelium and subventricular zone intermingled with the callosal glioepithelium

Paracentral (agranular/granular interface) stratified transitional field (STF)

STF1 t2
STF2
STF2/3
STF4/3
STF4 t2
STF5?
STF6

PARACENTRAL LOBULE

Subpial granular layer (glioepithelium)
Layer I
Cortical plate
Layer VII (subplate)

Postcentral gyrus

Lateral ventricle

Cingulate gyrus

Cingulum

Fornical glioepithelium

Callosal glioepithelium

Callosal sling?

Lateral fissure

Internal capsule

Corpus callosum

Caudate nucleus (body)

Induseum griseum

Choroid plexus

Paracentral neuroepithelium and subventricular zone

Cortical/striatal neuroepithelium and subventricular zone

Posterior striatal neuroepithelium and subventricular zone

Strionuclear glioepithelium

Fornix

Dorsolateral nucleus

Stria terminalis

Stria medullaris

Choroid plexus

Superior temporal gyrus

External capsule

Claustrum?

Ventral lateral nucleus

Dorsomedial nucleus

Third ventricle

Dorsal complex

THALAMUS

Reticular belt

TEMPORAL LOBE

BASAL GANGLIA

Internal capsule

Reticular nucleus

Ventral posterolateral nucleus

Thalamic glioepithelium/ ependyma

Ventral complex

Strionuclear glioepithelium

Posterior striatal neuroepithelium and subventricular zone

Cortical/striatal neuroepithelium and subventricular zone

Periventricular complex

Putamen

Globus pallidus

Ventral posteromedial nucleus

Zona incerta

Reuniens nucleus

Junction between strionuclear and perifascicular glioepithelia

Temporal neuroepithelium and subventricular zone

Internal capsule (posterior limb)

Subthalamic nucleus

Third ventricle

Perifascicular glioepithelium

Anterior commissure?

Stria terminalis

Optic tract

Forel's fields

Mammillo-thalamic tract

Subthalamic glioepithelium/ependyma

STF4 t1

Caudate nucleus (tail)

Cerebral peduncle

Fornix

Supra-mammillary area

Hypothalamic glioepithelium/ependyma

Medial forebrain bundle

Mammillary body

HYPO-THALAMUS

Dentate migration

Dentate gyrus

Ammon's horn

Medial mammillary nucleus

Subgranular zone

HIPPO-CAMPUS

Ammonic migration and sojourn zone

Lateral ventricle

Lateral mammillary nucleus

Third ventricle (mammillary recess)

Hippocampal neuroepithelium

Entorhinal cortex

Parahippocampal neuroepithelium and subventricular zone

STF3abc
STF2
STF4 t2
STF5
STF1 t2

Parahippocampal gyrus

Temporal (granular) stratified transitional field (STF)

Germinal and transitional structures in *italics*

56

PLATE 27A
CR 165 mm
GW 20, Y17-63
Frontal
Section 511

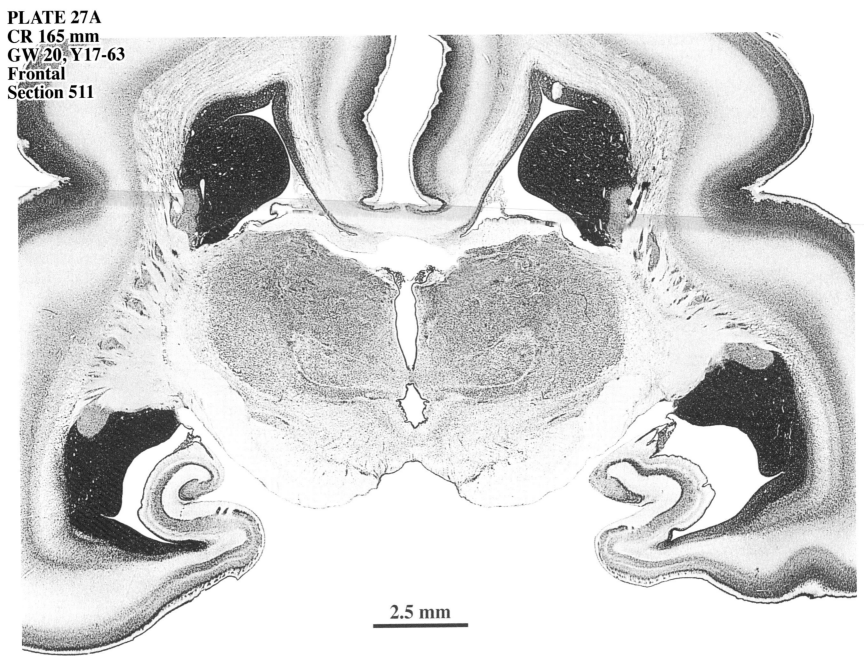

2.5 mm

See the entire section in Plates 15A and B.

PARACENTRAL LOBULE

Interhemispheric fissure

Cingulate neuroepithelium and subventricular zone intermingled with the callosal glioepithelium

Subpial granular layer (glioepithelium)
Layer I
Cortical plate
Layer VII (subplate)

Postcentral gyrus

Paracentral (agranular/granular interface) stratified transitional field (STF)

STF1 t2
STF2
STF3a
STF3b
STF4 t2
STF4/3
STF6

Lateral ventricle

Fornical glioepithelium

Lateral fissure

Cingulate gyrus

Induseum griseum

Callosal glioepithelium

Paracentral neuroepithelium and subventricular zone

Superior temporal gyrus

Caudate nucleus (tail)

Fornix

Corpus callosum

Cingulum

Callosal sling?

Cortical/striatal neuroepithelium and subventricular zone

Posterior striatal neuroepithelium and subventricular zone

Strionuclear glioepithelium

Choroid plexus

Dorsolateral nucleus

Stria terminalis

Internal capsule

Dorsal complex

THALAMUS

Strionuclear glioepithelium

Posterior striatal neuroepithelium and subventricular zone

Putamen (islands)

Reticular nucleus

Ventral lateral nucleus

Stria medullaris

Cortical/striatal neuroepithelium and subventricular zone

Dorsomedial nucleus

Thalamic glioepithelium/ependyma

Ventral complex

Reticular belt

Temporal neuroepithelium and subventricular zone

TEMPORAL LOBE

Ventral posterolateral nucleus

Centromedian nucleus

Third ventricle

Periventricular complex

Central complex

Junction between strionuclear and perifascicular glioepithelia

Internal capsule (posterior limb)

Optic tract

Ventral posteromedial nucleus

Aqueduct

Caudate nucleus (tail)

Stria terminalis

Zona incerta

MIDBRAIN TEGMENTUM

Habenulo-interpeduncular tract

Mesencephalic glioepithelium/ependyma

Perifascicular glioepithelium

Medial forebrain bundle

Substantia nigra

Interpeduncular nucleus

Ammonic migration and sojourn zone

Cerebral peduncle

Dentate gyrus

Pars compacta

Pars reticulata

Lateral ventricle

Nerve III (oculomotor)

Ammon's horn

HIPPO-CAMPUS

Ventral tegmental area

Dentate migration

Subgranular zone

Inferior temporal gyrus

Entorhinal cortex

Parahippocampal neuroepithelium and subventricular zone

STF3abc
STF2
STF4 t2
STF5
STF1 t2

Parahippocampal gyrus

Hippocampal neuroepithelium

Temporal (granular) stratified transitional field (STF)

PLATE 28A
CR 165 mm
GW 20, Y17-63
Frontal
Section 551

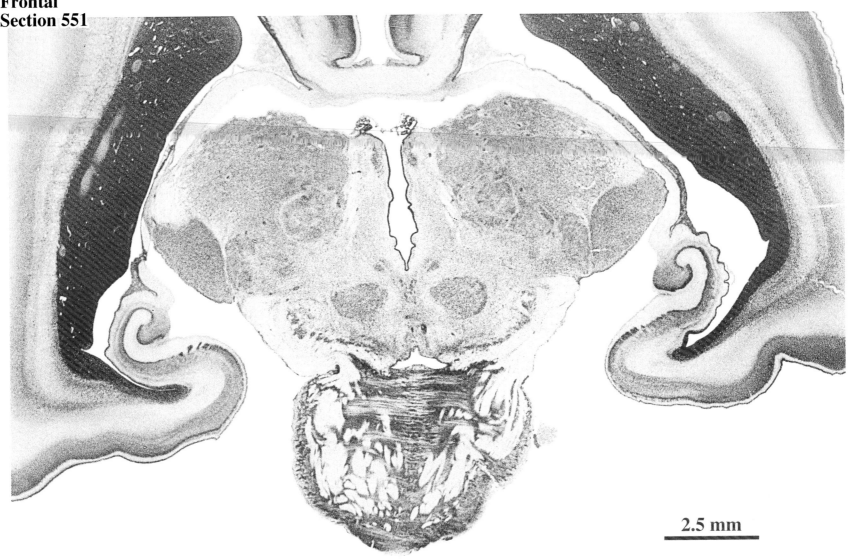

2.5 mm

See the entire section in Plates 16A and B.

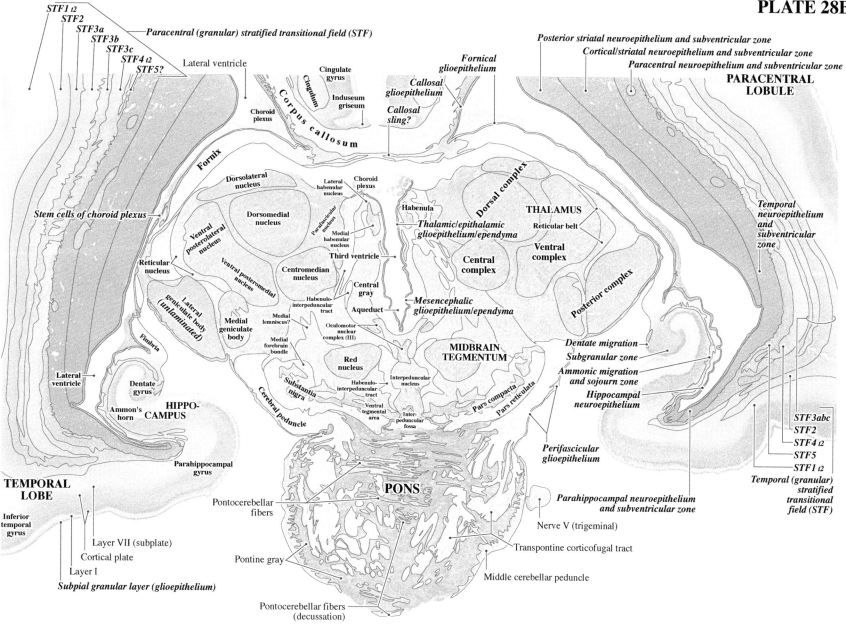

STF1 t2
STF2
STF3a
STF3b
STF3c
STF4 t2
STF5?

Paracentral (granular) stratified transitional field (STF)

Lateral ventricle

Cingulate gyrus

Cingulum

Induseum griseum

Corpus callosum

Choroid plexus

Fornical glioepithelium

Callosal glioepithelium

Callosal sling?

Posterior striatal neuroepithelium and subventricular zone
Cortical/striatal neuroepithelium and subventricular zone
Paracentral neuroepithelium and subventricular zone

PARACENTRAL LOBULE

Fornix

Dorsolateral nucleus

Lateral habenular nucleus

Choroid plexus

Habenula

Thalamic/epithalamic glioepithelium/ependyma

THALAMUS

Reticular belt

Dorsal complex

Temporal neuroepithelium and subventricular zone

Stem cells of choroid plexus

Dorsomedial nucleus

Ventral posterolateral nucleus

Parafascicular nucleus

Medial habenular nucleus

Third ventricle

Ventral complex

Reticular nucleus

Ventral posteromedial nucleus

Centromedian nucleus

Central gray

Central complex

Posterior complex

Lateral geniculate body (unlaminated)

Medial geniculate body

Habenulo-interpeduncular tract

Central gray

Aqueduct

Mesencephalic glioepithelium/ependyma

Fimbria

Medial lemniscus?

Medial forebrain bundle

Oculomotor nuclear complex (III)

MIDBRAIN TEGMENTUM

Dentate migration
Subgranular zone

Lateral ventricle

Dentate gyrus

Red nucleus

Substantia nigra

Habenulo-interpeduncular tract

Interpeduncular nucleus

Ammonic migration and sojourn zone

Ammon's horn

HIPPO-CAMPUS

Cerebral peduncle

Ventral tegmental area

Inter-peduncular fossa

Pars compacta

Pars reticulata

Hippocampal neuroepithelium

Perifascicular glioepithelium

STF3abc
STF2
STF4 t2
STF5
STF1 t2

Parahippocampal gyrus

TEMPORAL LOBE

PONS

Parahippocampal neuroepithelium and subventricular zone

Temporal (granular) stratified transitional field (STF)

Inferior temporal gyrus

Pontocerebellar fibers

Nerve V (trigeminal)

Transpontine corticofugal tract

Layer VII (subplate)
Cortical plate
Layer I

Pontine gray

Middle cerebellar peduncle

Subpial granular layer (glioepithelium)

Pontocerebellar fibers (decussation)

Germinal and transitional structures in *italics*

PLATE 29A
CR 165 mm
GW 20, Y17-63
Frontal
Section 601

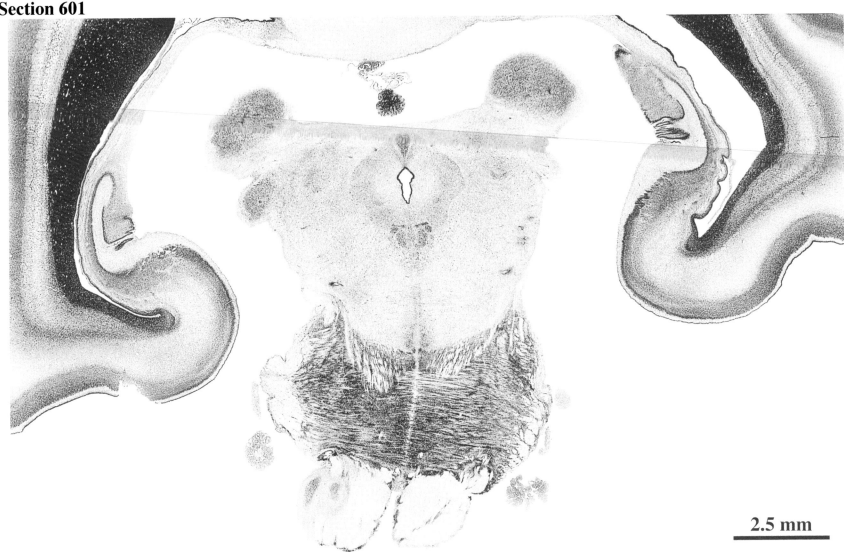

2.5 mm

See the entire section in Plates 17A and B.

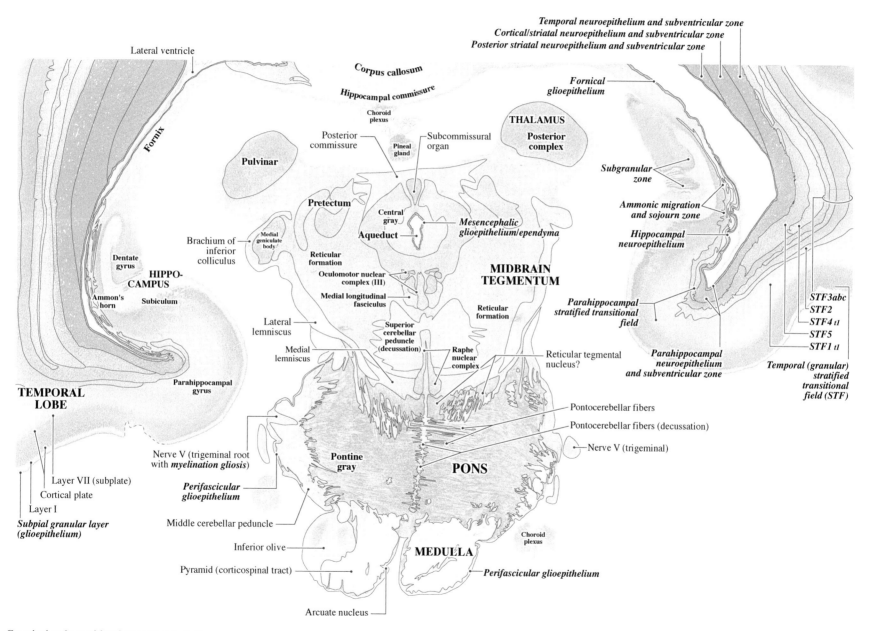

Germinal and transitional structures in *italics*

PLATE 30A
CR 165 mm
GW 20, Y17-63
Frontal
Section 631

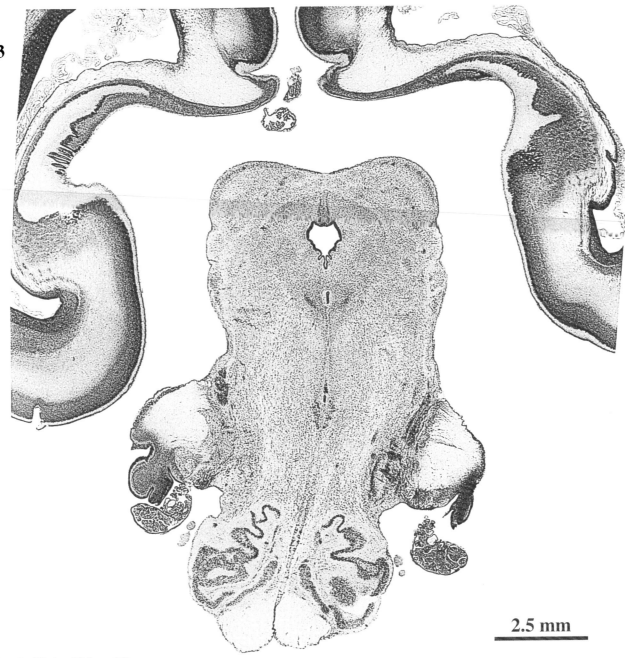

2.5 mm

See the entire section in Plates 18A and B.

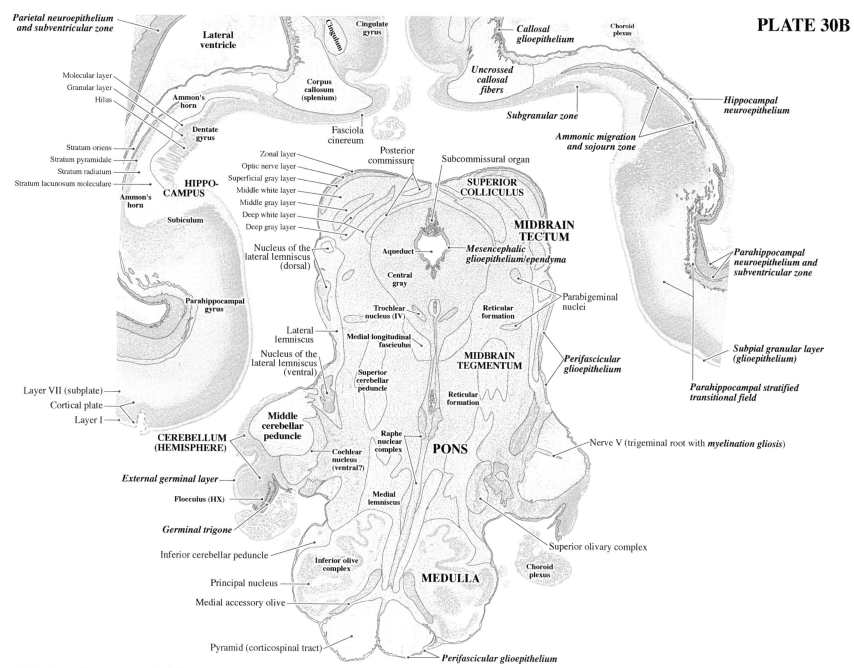

Parietal neuroepithelium and subventricular zone

Lateral ventricle

Cingulate gyrus

Callosal glioepithelium

Choroid plexus

Cingulum

Uncrossed callosal fibers

Hippocampal neuroepithelium

Molecular layer

Granular layer

Hilus

Ammon's horn

Corpus callosum (splenium)

Subgranular zone

Dentate gyrus

Fasciola cinereum

Ammonic migration and sojourn zone

Stratum oriens

Stratum pyramidale

Stratum radiatum

Stratum lacunosum moleculare

Zonal layer

Optic nerve layer

Superficial gray layer

Middle white layer

Middle gray layer

Deep white layer

Deep gray layer

Posterior commissure

Subcommissural organ

SUPERIOR COLLICULUS

HIPPO-CAMPUS

Ammon's horn

Subiculum

MIDBRAIN TECTUM

Parahippocampal neuroepithelium and subventricular zone

Nucleus of the lateral lemniscus (dorsal)

Aqueduct

Mesencephalic glioepithelium/ependyma

Central gray

Parahippocampal gyrus

Trochlear nucleus (IV)

Reticular formation

Parabigeminal nuclei

Lateral lemniscus

Medial longitudinal fasciculus

Nucleus of the lateral lemniscus (ventral)

MIDBRAIN TEGMENTUM

Perifascicular glioepithelium

Subpial granular layer (glioepithelium)

Superior cerebellar peduncle

Layer VII (subplate)

Cortical plate

Layer I

Reticular formation

Parahippocampal stratified transitional field

Middle cerebellar peduncle

Raphe nuclear complex

CEREBELLUM (HEMISPHERE)

PONS

Nerve V (trigeminal root with *myelination gliosis*)

External germinal layer

Cochlear nucleus (ventral?)

Flocculus (HX)

Medial lemniscus

Germinal trigone

Inferior cerebellar peduncle

Superior olivary complex

Inferior olive complex

Choroid plexus

Principal nucleus

MEDULLA

Medial accessory olive

Pyramid (corticospinal tract)

Perifascicular glioepithelium

Germinal and transitional structures in *italics*

64

PLATE 31A
CR 165 mm
GW 20, Y17-63
Frontal
Section 661

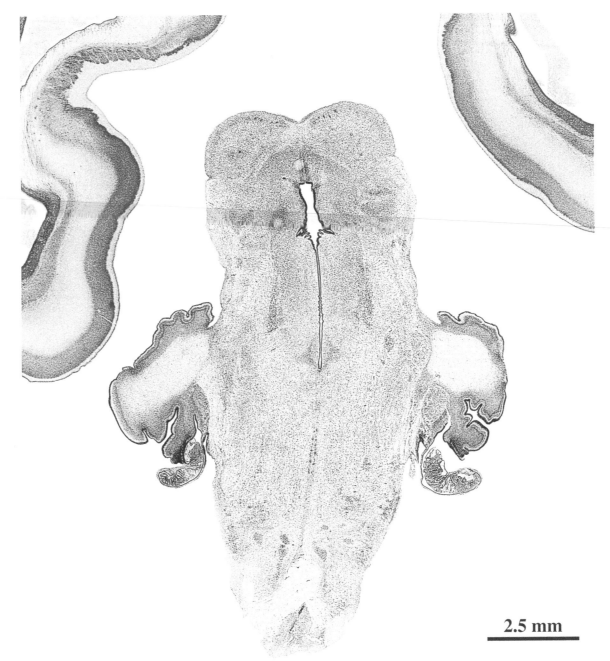

2.5 mm

See the entire section in Plates 19A and B.

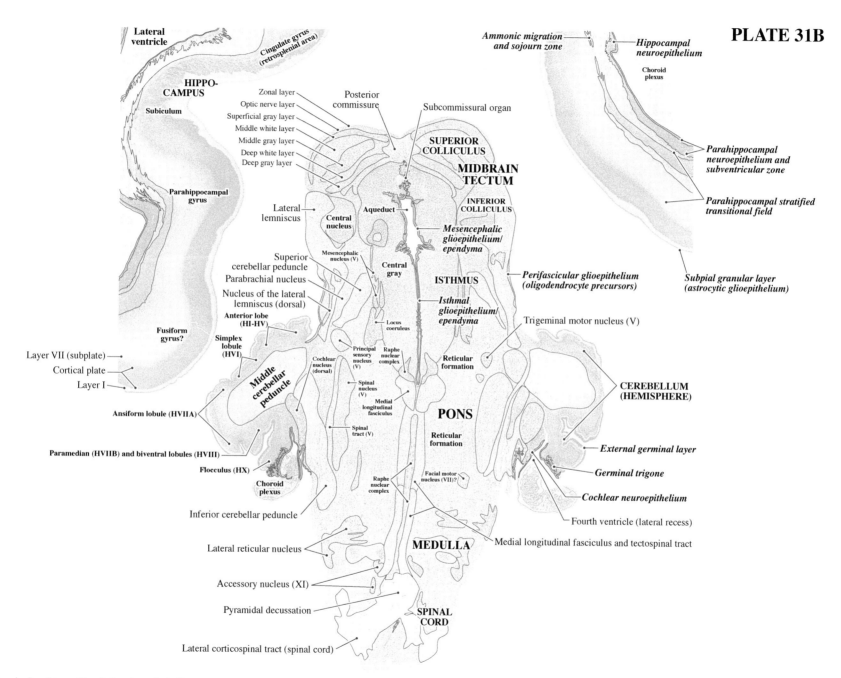

65

PLATE 31B

Germinal and transitional structures in *italics*

PLATE 32A
CR 165 mm
GW 20, Y17-63
Frontal
Section 701

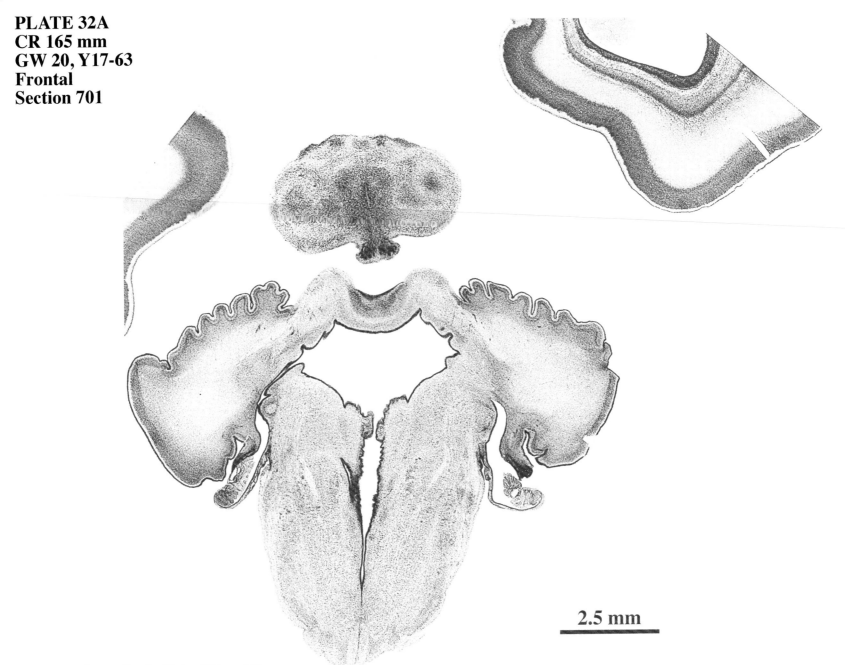

2.5 mm

See the entire section in Plates 20A and B.

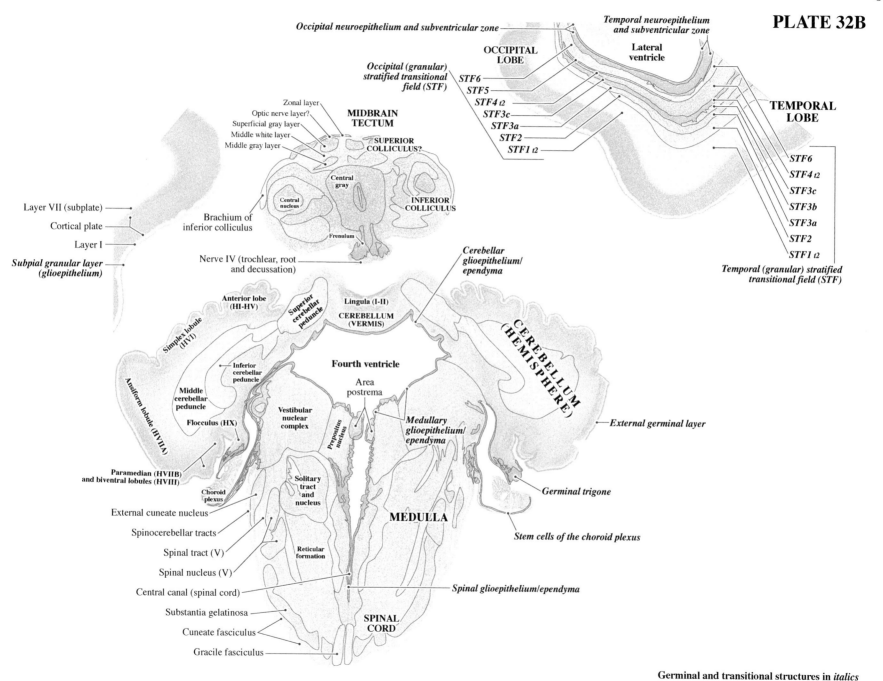

Occipital neuroepithelium and subventricular zone

Temporal neuroepithelium and subventricular zone

OCCIPITAL LOBE

Lateral ventricle

TEMPORAL LOBE

Occipital (granular) stratified transitional field (STF)

STF6
STF5
STF4 t2
STF3c
STF3a
STF2
STF1 t2

STF6
STF4 t2
STF3c
STF3b
STF3a
STF2
STF1 t2

Temporal (granular) stratified transitional field (STF)

Zonal layer
Optic nerve layer?
Superficial gray layer
Middle white layer
Middle gray layer

MIDBRAIN TECTUM

SUPERIOR COLLICULUS?

Central gray

Central nucleus

INFERIOR COLLICULUS

Layer VII (subplate)
Cortical plate
Layer I
Subpial granular layer (glioepithelium)

Brachium of inferior colliculus

Frenulum

Nerve IV (trochlear, root and decussation)

Cerebellar glioepithelium/ ependyma

Anterior lobe (HI-HV)

Superior cerebellar peduncle

Lingula (I-II)

CEREBELLUM (VERMIS)

Simplex lobule (HVI)

CEREBELLUM (HEMISPHERE)

Ansiform lobule (HVIIA)

Inferior cerebellar peduncle

Fourth ventricle

Area postrema

Middle cerebellar peduncle

Flocculus (HX)

Vestibular nuclear complex

Prepositus nucleus

Medullary glioepithelium/ ependyma

External germinal layer

Paramedian (HVIIB) and biventral lobules (HVIII)

Choroid plexus

External cuneate nucleus

Spinocerebellar tracts

Spinal tract (V)

Spinal nucleus (V)

Central canal (spinal cord)

Substantia gelatinosa

Cuneate fasciculus

Gracile fasciculus

Solitary tract and nucleus

MEDULLA

Reticular formation

Germinal trigone

Stem cells of the choroid plexus

Spinal glioepithelium/ependyma

SPINAL CORD

Germinal and transitional structures in *italics*

68

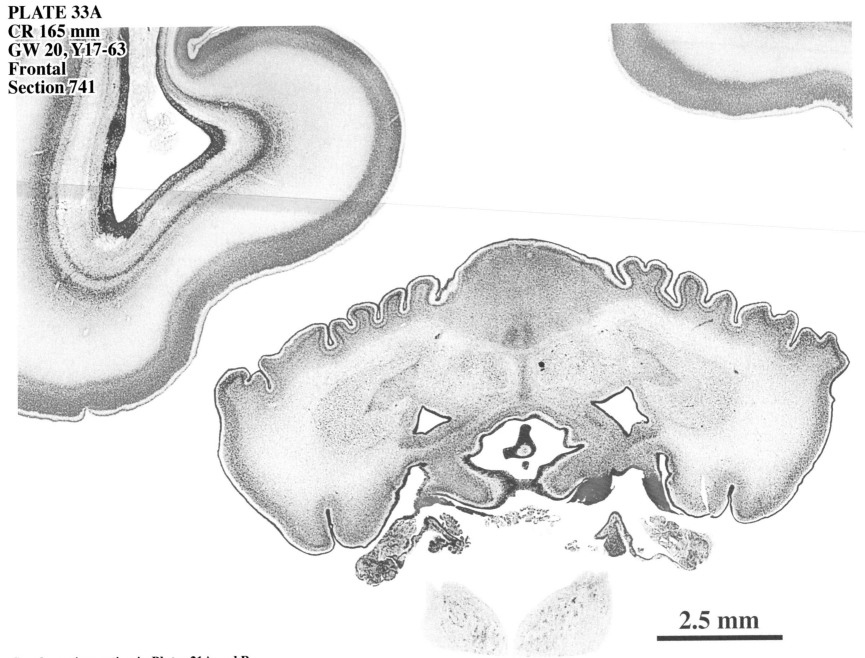

2.5 mm

See the entire section in Plates 21A and B.

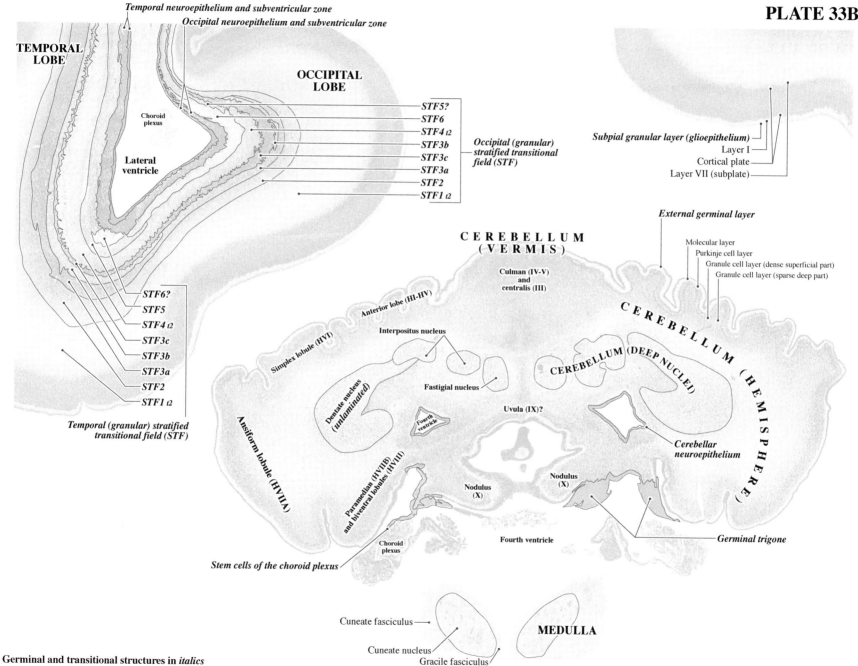

Temporal neuroepithelium and subventricular zone
Occipital neuroepithelium and subventricular zone

TEMPORAL LOBE

Choroid plexus

OCCIPITAL LOBE

Lateral ventricle

STF5?
STF6
STF4 *t2*
STF3b
STF3c
STF3a
STF2
STF1 *t2*

Occipital (granular) stratified transitional field (STF)

Subpial granular layer (glioepithelium)
Layer I
Cortical plate
Layer VII (subplate)

External germinal layer

Molecular layer
Purkinje cell layer
Granule cell layer (dense superficial part)
Granule cell layer (sparse deep part)

STF6?
STF5
STF4 *t2*
STF3c
STF3b
STF3a
STF2
STF1 *t2*

Temporal (granular) stratified transitional field (STF)

C E R E B E L L U M (V E R M I S)

Culman (IV-V) and centralis (III)

Anterior lobe (HI-HV)

Interpositus nucleus

C E R E B E L L U M   (H E M I S P H E R E)

Simplex lobule (HVI)

Fastigial nucleus

CEREBELLUM (DEEP NUCLEI)

Dentate nucleus (*unlaminated*)

Fourth ventricle

Uvula (IX)?

Ansiform lobule (HVIIA)

Paramedian (HVIIB) and biventral lobules (HVIII)

Nodulus (X)

Nodulus (X)

Cerebellar neuroepithelium

Fourth ventricle

Choroid plexus

Germinal trigone

Stem cells of the choroid plexus

Cuneate fasciculus

MEDULLA

Cuneate nucleus
Gracile fasciculus

Germinal and transitional structures in *italics*

PLATE 34A
CR 165 mm
GW 20, Y17-63
Frontal, Section 821

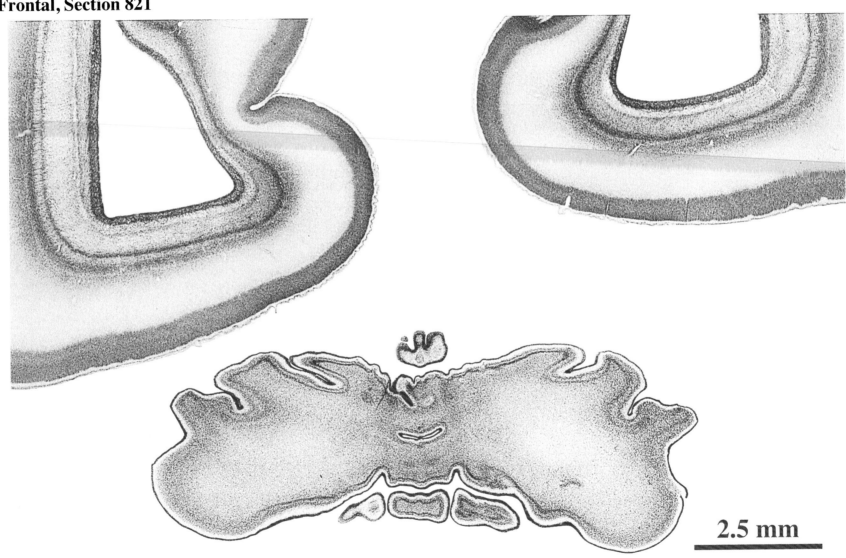

2.5 mm

See the entire section in Plates 22A and B.

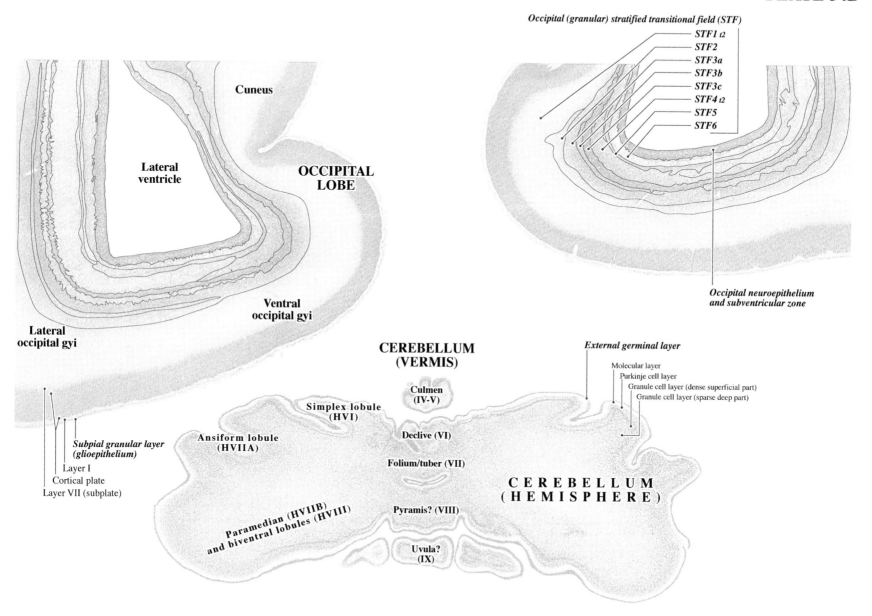

Occipital (granular) stratified transitional field (STF)

STF1 t2
STF2
STF3a
STF3b
STF3c
STF4 t2
STF5
STF6

Cuneus

Lateral ventricle

OCCIPITAL LOBE

Occipital neuroepithelium and subventricular zone

Ventral occipital gyi

Lateral occipital gyi

Subpial granular layer (glioepithelium)

Layer I
Cortical plate
Layer VII (subplate)

CEREBELLUM (VERMIS)

External germinal layer

Molecular layer
Purkinje cell layer
Granule cell layer (dense superficial part)
Granule cell layer (sparse deep part)

Culmen (IV-V)

Simplex lobule (HVI)

Ansiform lobule (HVIIA)

Declive (VI)

Folium/tuber (VII)

C E R E B E L L U M (H E M I S P H E R E)

Paramedian (HVIIB) and biventral lobules (HVIII)

Pyramis? (VIII)

Uvula? (IX?)

Germinal and transitional structures in *italics*

**PLATE 35
CR 165 mm
GW 20, Y17-63
Frontal
Section 341
FRONTAL
CORTEX**

See the entire section 361 in Plates 12A and B.

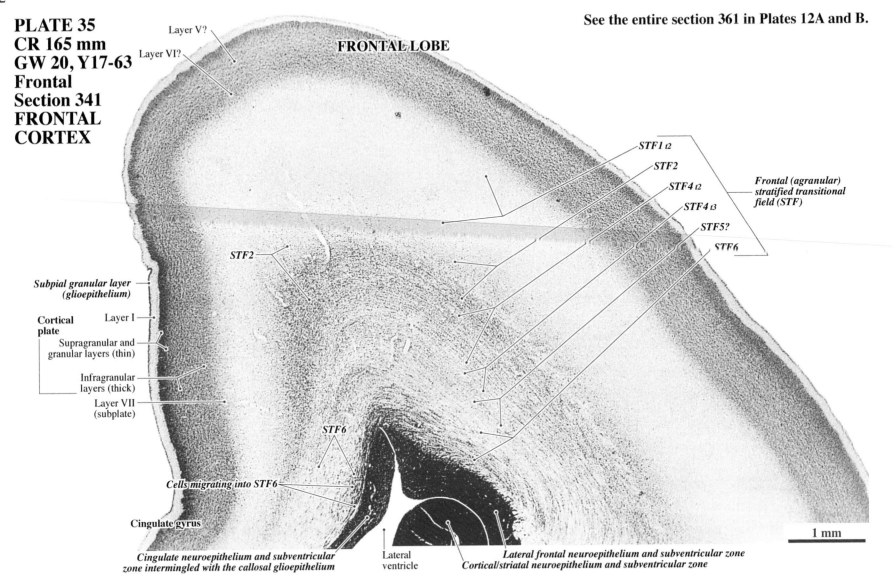

Layer V?

Layer VI?

FRONTAL LOBE

STF1 t2

STF2

STF4 t2

STF4 t3

STF5?

STF6

*Frontal (agranular)
stratified transitional
field (STF)*

STF2

*Subpial granular layer
(glioepithelium)*

**Cortical
plate**

Layer I

Supragranular and
granular layers (thin)

Infragranular
layers (thick)

Layer VII
(subplate)

STF6

Cells migrating into STF6

Cingulate gyrus

*Cingulate neuroepithelium and subventricular
zone intermingled with the callosal glioepithelium*

Lateral
ventricle

Lateral frontal neuroepithelium and subventricular zone
Cortical/striatal neuroepithelium and subventricular zone

1 mm

The dense superficial part of the **cortical plate** is presumed to be composed of recently arrived Layers IV (granular), III, and II (supragranular) neurons. The less dense deep portion of the cortical plate is composed mainly of Layers V and VI infragranular neurons. Layer V may be the slightly lighter band in the central cortical plate, and Layer VI may be the slightly darker band at the base. In this section ,the supragranular layers are thin and the infragranular layers are thick. That is because many of the neurons that will occupy the supragranular layers are still migrating and sojourning in the *stratified transitional field (STF).* As in all second trimester specimens, the thickest part of the cortex is the *STF* between Layer VII and the dense germinal matrix. In this area of agranular (sparse Layer IV) fron tal cortex, 5 of the 6 *STF* layers are distinguishable. *STF3* is absent in agranular cortical areas.

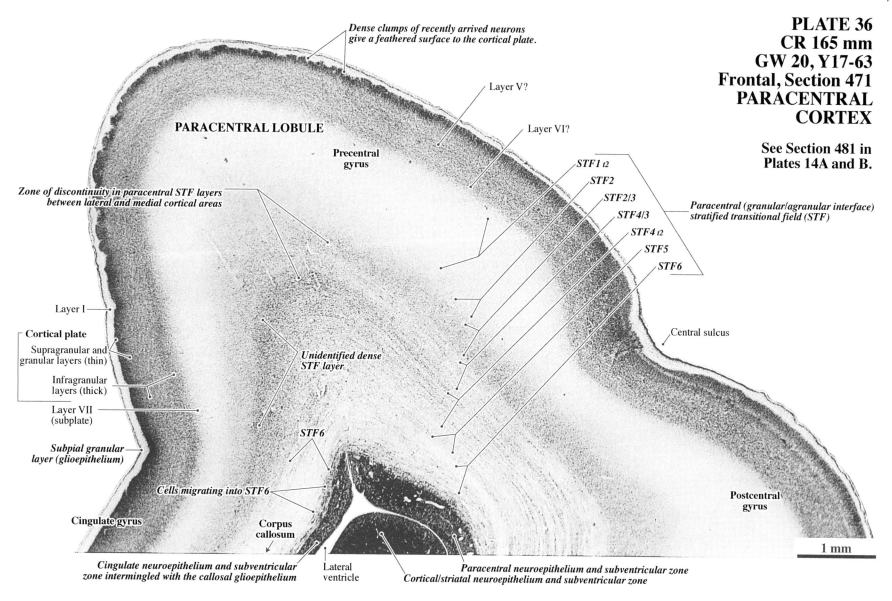

**PLATE 36
CR 165 mm
GW 20, Y17-63
Frontal, Section 471
PARACENTRAL
CORTEX**

See Section 481 in
Plates 14A and B.

Dense clumps of recently arrived neurons give a feathered surface to the cortical plate.

PARACENTRAL LOBULE

Layer V?

Layer VI?

Precentral gyrus

STF1 t2
STF2
STF2/3
STF4/3
STF4 t2
STF5
STF6

Paracentral (granular/agranular interface) stratified transitional field (STF)

Zone of discontinuity in paracentral STF layers between lateral and medial cortical areas

Layer I

Cortical plate
Supragranular and granular layers (thin)

Infragranular layers (thick)

Layer VII (subplate)

Subpial granular layer (glioepithelium)

Unidentified dense STF layer

Central sulcus

STF6

Cells migrating into STF6

Cingulate gyrus

Corpus callosum

Postcentral gyrus

1 mm

Cingulate neuroepithelium and subventricular zone intermingled with the callosal glioepithelium

Lateral ventricle

Paracentral neuroepithelium and subventricular zone
Cortical/striatal neuroepithelium and subventricular zone

This part of the **cortical plate** has a thick infragranular part and a thin supragranular part medial and dorsal to the central sulcus. Near the central sulcus, Layer V may be the dense middle band, and Layer VI may be the dense deep band. The surface is feathered in some areas of the paracentral cortex. That is seen in many specimens and is probably caused by the massive accumulations of young neurons that have just migrated to the cortical plate. Layers of the *stratified transitional field (STF)* are beginning to take on some of the characteristics of granular cortex in this transition area. *STF3* appears as a cell-dense layer blending with *STF2* and a fibrous layer blending with *STF4*. The dense *STF2/3* layer does not continue into the medial wall of the cortex. Instead, there is an unidentified deeper dense layer that begins in the dorsomedial cortex and continues into the cingulate gyrus. It is possible that the *discontinuity* between the medial and lateral *STF* layers represents differences between the limbic cingulate gyrus and the neocortical paracentral lobule.

74

PLATE 37
CR 165 mm
GW 20, Y17-63
Frontal
Section 731
PARIETAL
CORTEX

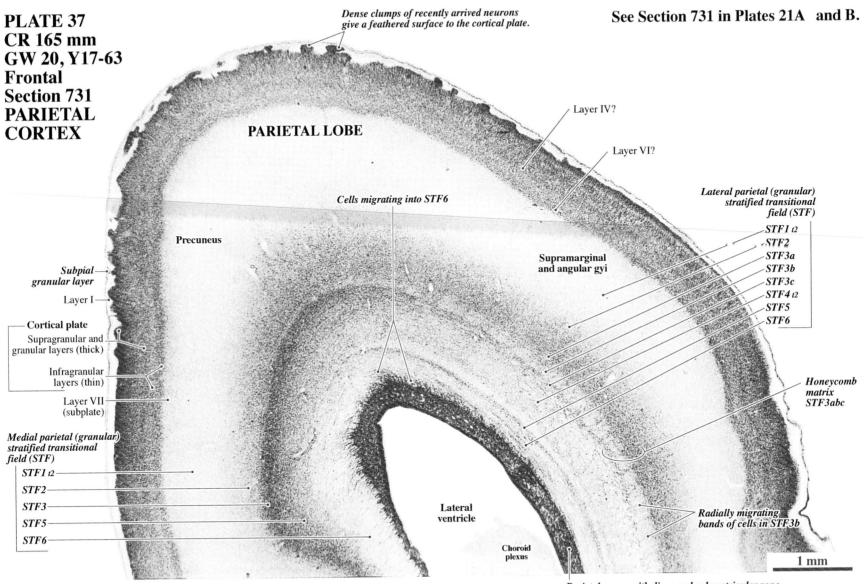

*Dense clumps of recently arrived neurons
give a feathered surface to the cortical plate.*

PARIETAL LOBE

Layer IV?

Layer VI?

*Lateral parietal (granular)
stratified transitional
field (STF)*

Cells migrating into STF6

Precuneus

STF1 t2
STF2
STF3a
STF3b
STF3c
STF4 t2
STF5
STF6

Supramarginal
and angular gyi

*Subpial
granular layer*

Layer I

Cortical plate
Supragranular and
granular layers (thick)

Infragranular
layers (thin)

Layer VII
(subplate)

*Honeycomb
matrix
STF3abc*

*Medial parietal (granular)
stratified transitional
field (STF)*

STF1 t2

STF2

STF3

STF5

STF6

Lateral
ventricle

*Radially migrating
bands of cells in STF3b*

Choroid
plexus

1 mm

Parietal neuroepithelium and subventricular zone

The **cortical plate** in this part of the parietal cortex has thicker granular/supragranular layers than infragranular layers. Layer IV will be prominent in this part of the cortex when it is mature. Layer IV may be recognized as the dense middle band in the cortical plate, and Layer VI may be the dense deep band. Layers in the *stratified transitional field (STF)* are characteristic of granular (future sensory) cortex. The most notable characteristic of the lateral parietal *STF* is the differentiation of *STF3* into a *honeycomb matrix* with three sublayers: *STF3a* is a dense cellular band that blends with *STF2*; *STF3b* is a fibrous band with fine vertical striations created by strands of cells migrating radially through the fibers; *STF3c* is a thin cellular band above *STF4*. In the medial parietal cortex, *STF2* and *STF6* are similar in thickness. *STF5* is very thick medially, and *STF4* is absent. Differentiation of *STF3* into its three sublayers has not yet developed in the medial parietal cortex.

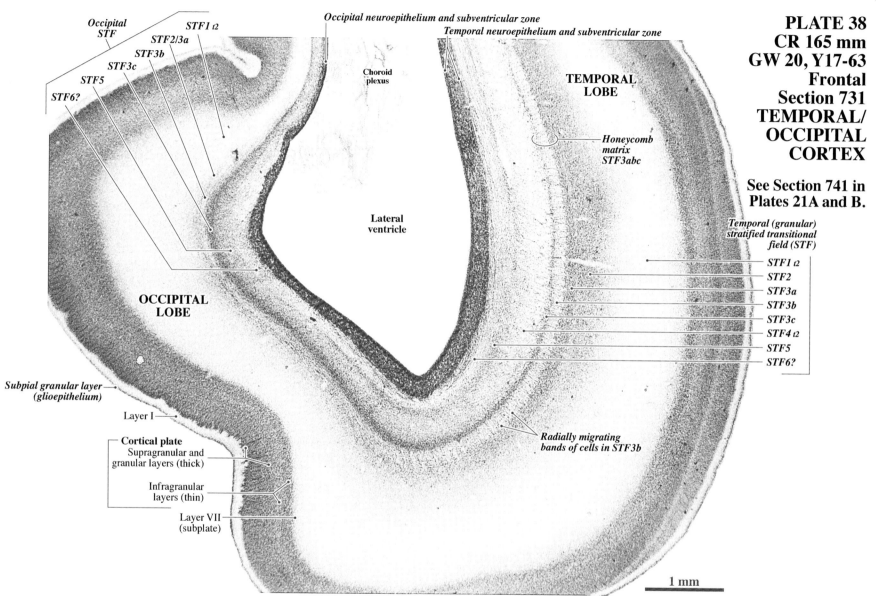

PLATE 38
CR 165 mm
GW 20, Y17-63
Frontal
Section 731
TEMPORAL/
OCCIPITAL
CORTEX

See Section 741 in
Plates 21A and B.

Occipital neuroepithelium and subventricular zone

Temporal neuroepithelium and subventricular zone

*Occipital
STF*

STF1 *t2*

STF2/3a

STF3b

STF3c

STF5

STF6?

Choroid
plexus

Lateral
ventricle

**TEMPORAL
LOBE**

*Honeycomb
matrix
STF3abc*

*Temporal (granular)
stratified transitional
field (STF)*

STF1 *t2*

STF2

STF3a

STF3b

STF3c

STF4 *t2*

STF5

STF6?

**OCCIPITAL
LOBE**

*Subpial granular layer
(glioepithelium)*

Layer I

Cortical plate
Supragranular and
granular layers (thick)

Infragranular
layers (thin)

Layer VII
(subplate)

*Radially migrating
bands of cells in STF3b*

1 mm

The **cortical plate** in both the temporal and occipital cortices has thicker granular/supragranular parts than infragranular parts because these areas will have a dense Layer IV at maturity. The *STF3 honeycomb matrix* is well differentiated in the lateral temporal cortex.

PLATE 39
CR 165 mm, GW20, Y17-63, Frontal

1. Rostral-to-caudal Gradient in Crossing of the Corpus Callosum

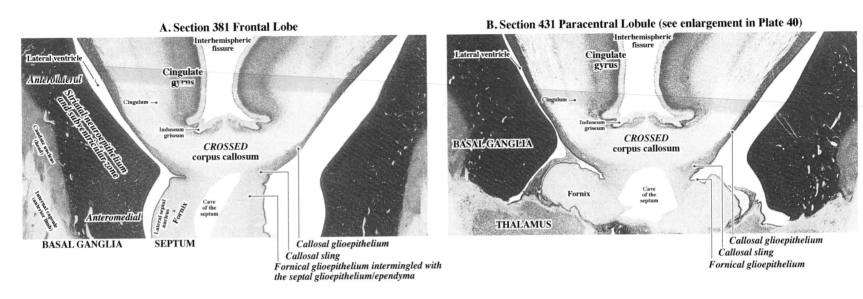

A. Section 381 Frontal Lobe

B. Section 431 Paracentral Lobule (see enlargement in Plate 40)

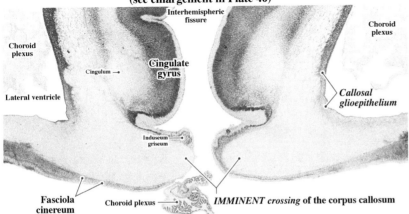

C. Section 621 Paracentral Lobule/Parietal Lobe Interface
(see enlargement in Plate 40)

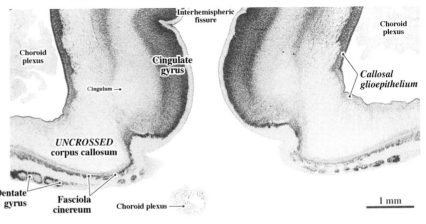

D. Section 641 Paracentral Lobule/Parietal Lobe Interface

1 mm

PLATE 40
CR 165 mm, GW20, Y17-63, Frontal

2. Rostral-to-caudal Gradient in Crossing of the Corpus Callosum

A. Section 431
Paracentral Lobule

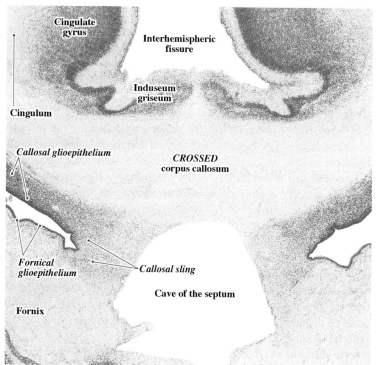

Cingulate gyrus

Interhemispheric fissure

Induseum griseum

Cingulum

Callosal glioepithelium

CROSSED corpus callosum

Fornical glioepithelium

Callosal sling

Cave of the septum

Fornix

B. Section 621
Paracentral Lobule/Parietal Lobe Interface

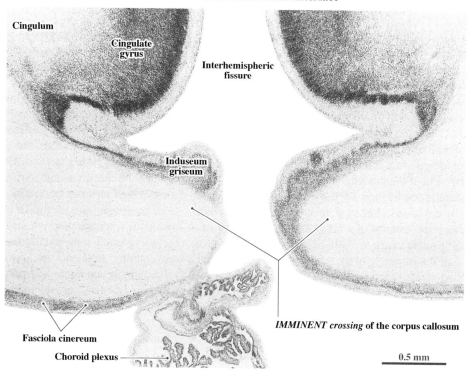

Cingulum

Cingulate gyrus

Interhemispheric fissure

Induseum griseum

Fasciola cinereum

Choroid plexus

IMMINENT crossing of the corpus callosum

0.5 mm

PART IV: Y391-62
CR 170 mm (GW 20)
Horizontal

This specimen is a stillborn female (Yakovlev case number RP3L-B 391 62, referred to as Y391-62) with a crown-rump length (CR) of 170 mm and estimated to be at gestational week (GW) 20. The brain was cut in the horizontal plane in 690 sections (35-μm thick) and is classified as a Normative Control in the Yakovlev Collection (Haleem, 1990). Since there is no photograph of this brain before it was embedded and cut, we turned to the comprehensive atlas that Retzius published in 1896 showing whole fetal brains in medial, lateral, superior, and inferior views and midline sagittally cut brains. **Figure 5**, taken from Retzius (1896), shows the exterior of a brain from a specimen that is comparable in age to Y391-62, along with the approximate cutting angle of the sections. Photographs of 12 Nissl-stained sections are illustrated in **Plates 41–48** and show low-magnification photos that contain the cortex as well as the brainstem. **Plates 49–59** show high-magnification views of the brain core.

The ***cortical neuroepithelium and subventricular zone*** is generating neocortical neurons mainly for superficial layers. The ***stratified transitional fields*** in all lobes of the cerebral cortex are filled with migrating and sojourning neurons and show distinct regional heterogeneity between granular (future sensory) and agranular (future motor) areas. The cerebral cortex is smooth except for the lateral fissure, the cingulate sulcus, the parietal occipital sulcus, and the calcarine sulcus; the narrow invaginations seen in anterior and posterior regions are shrinkage artifacts produced during fixation. Many neurons, glia, and their mitotic precursor cells are still migrating through the olfactory peduncle toward the olfactory bulb (***rostral migratory stream***) from a presumed source area in the germinal matrix at the junction between the cerebral cortex and striatum (***cortical/striatal neuroepithelium and subventricular zone***). In anterolateral parts of the cerebral cortex, streams of neurons and glia are numerous in the ***lateral migratory stream***. That stream percolates through the claustrum, endopiriform nucleus, external capsule, and uncinate fasciculus, and the cells appear to be heading toward the insular cortex, primary olfactory cortex, temporal cortex, and basolateral parts of the amygdaloid complex. In the hippocampus, cells are entering Ammon's horn pyramidal layer via a more definite ***ammonic migration***, and granule cells and their precursors are migrating to the dentate gyrus in the ***dentate migration***. There is a definite granule cell layer in the dentate gyrus. Stem cells of dentate granule cells populate the periphery of the dentate hilus just beneath the granular layer to form the ***subgranular zone*** where dentate granule cells are generated. A massive ***neuroepithelium/subventricular zone*** is in the nucleus accumbens and striatum where neurons (and glia) are being generated. The ***striatal neuroepithelium and subventricular zone*** has indistinct subdivisions, but the ***strionuclear glioepithelium*** forms a definite bulge at its medial edge. That glioepithelium may provide glia for the stria terminalis, internal capsule, stria medullaris, and contribute glial precursors to the ***perifascicular glioepithelium*** bordering the optic tract and the cerebral peduncle. There is a thin ***hippocampal neuroepithelium*** that blends with the ***fornical glioepithelium***. The septum has a ***glioepithelium/ependyma*** at the ventricular surface.

Neurons in most diencephalic structures appear to be settled; the major exceptions are the immature appearance of the lateral and medial geniculate bodies in the posterior thalamus and the hypothalamic medial mammillary body. The third ventricle is lined with a thin ***glioepithelium/ependyma*** that shows small invaginations and evaginations that mark previous sites where fate-restricted neuroepithelial patches generated neurons for specific nuclei. In the midbrain, pons, and medulla, there is a convoluted ***glioepithelium/ependyma*** lining the cerebral aqueduct and fourth ventricle. The pontine gray is enlarging and only a few neurons are still migrating into it from the ***anterior extramural migratory stream***.

The cerebellum is enlarging and lobules are identifiable in the vermis and the hemispheres. The deep nuclei are in place beneath the cortex. The entire surface of the cerebellar cortex is covered by the prominent ***external germinal layer (egl)*** that is actively producing basket, stellate, and granule cells. Lamination in the cortex is nearly absent, except for a thin molecular layer beneath the ***egl***. Nearly all Purkinje cells are migrating, a few are settling. The ***germinal trigone*** is large at the base of the nodulus and along the floccular peduncle.

GW20 HORIZONTAL SECTION PLANES

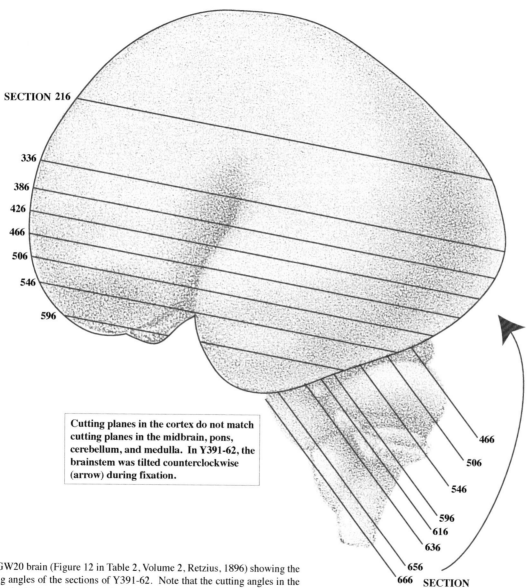

SECTION 216

336

386

426

466

506

546

596

Cutting planes in the cortex do not match cutting planes in the midbrain, pons, cerebellum, and medulla. In Y391-62, the brainstem was tilted counterclockwise (arrow) during fixation.

466

506

546

596

616

636

656

666 SECTION

Figure 5. The lateral view of a GW20 brain (Figure 12 in Table 2, Volume 2, Retzius, 1896) showing the approximate locations and cutting angles of the sections of Y391-62. Note that the cutting angles in the brainstem do no t match cutting angles in the cortex. Once the brain has been dissected from the skull, the brainstem has no support to maintain a constant angle of downward extension from the cortex. In Y391-62, the brainstem was flexed counterclockwise (arrow) to tuck under the cerebral hemispheres.

PLATE 41A
CR 170 mm
GW 20, Y391-62
Horizontal
Section 216

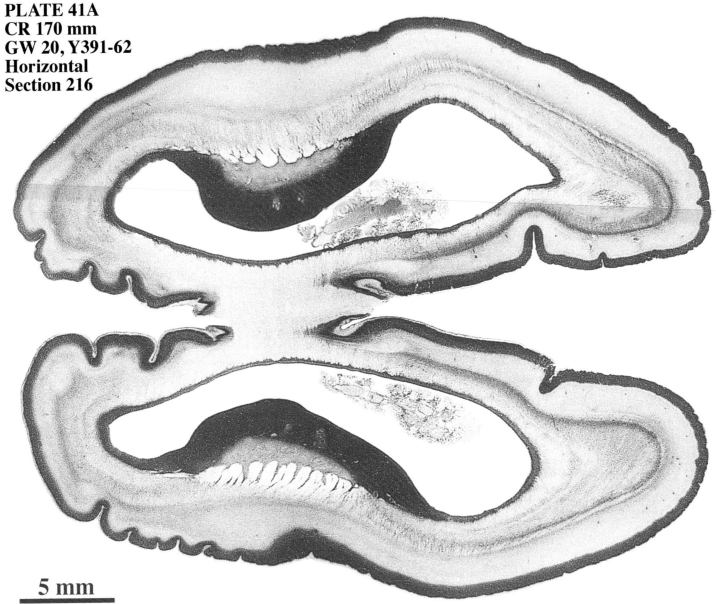

5 mm

LAYERS OF THE CORTICAL
STRATIFIED TRANSITIONAL
FIELD (STF)

STF1—Superficial fibrous layer with an early developmental stage *(t1)* when many cells are migrating through it, followed by a late stage *(t2)* with sparse cells. Endures as the subcortical white matter.

STF2—Upper cellular layer, the last sojourn zone before cells translocate to the cortical plate.

STF3—Honeycomb trilaminar matrix *(3a, 3b, 3c)* of cells and fibers found only in granular cortices.

STF4—Complex middle layer with three developmental stages:
t1– fibrous layer without interspersed cells;
t2– cells and fibers intermingle to form striations;
t3– fibers endure in the deep white matter.

STF5—Deep cellular layer, the first sojourn zone to appear outside the germinal matrix.

STF6—Late-forming deep layer of callosal fibers outside the germinal matrix.

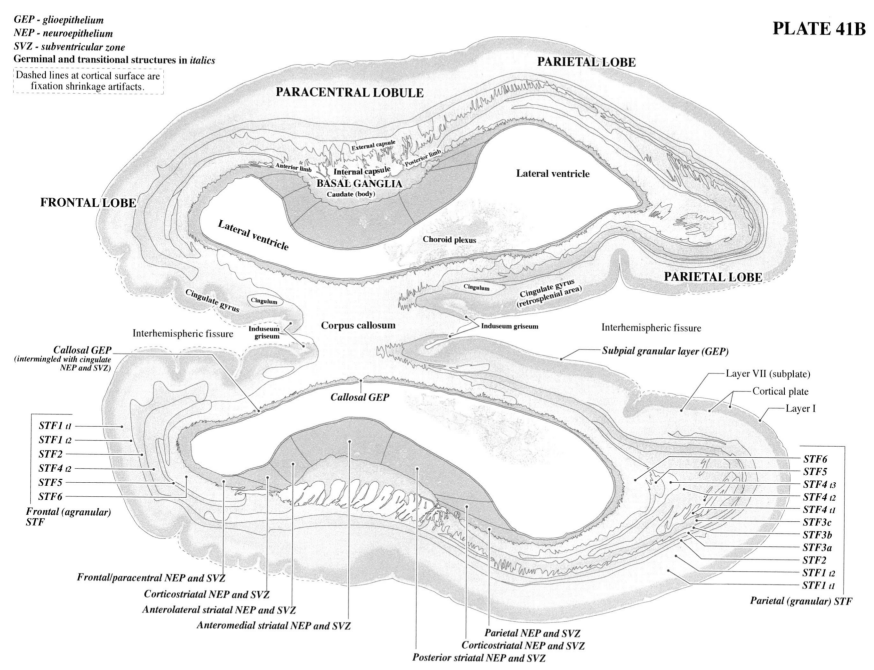

GEP - glioepithelium
NEP - neuroepithelium
SVZ - subventricular zone
Germinal and transitional structures in *italics*

Dashed lines at cortical surface are
fixation shrinkage artifacts.

PARACENTRAL LOBULE

PARIETAL LOBE

External capsule

Anterior limb   Internal capsule   Posterior limb

Lateral ventricle

BASAL GANGLIA
Caudate (body)

FRONTAL LOBE

Lateral ventricle

Choroid plexus

PARIETAL LOBE

Cingulate gyrus   Cingulum

Cingulum   Cingulate gyrus
(retrosplenial area)

Induseum griseum

Corpus callosum   *Induseum griseum*   Interhemispheric fissure

Interhemispheric fissure

Callosal GEP
(intermingled with cingulate NEP and SVZ)

Subpial granular layer (GEP)

Layer VII (subplate)

Cortical plate

Callosal GEP

Layer I

STF1 t1
STF1 t2
STF2
STF4 t2
STF5
STF6

STF6
STF5
STF4 t3
STF4 t2
STF4 t1
STF3c
STF3b
STF3a
STF2
STF1 t2
STF1 t1

Frontal (agranular) STF

Parietal (granular) STF

Frontal/paracentral NEP and SVZ

Corticostriatal NEP and SVZ

Anterolateral striatal NEP and SVZ

Anteromedial striatal NEP and SVZ

Parietal NEP and SVZ

Corticostriatal NEP and SVZ

Posterior striatal NEP and SVZ

PLATE 42A
CR 170 mm
GW 20, Y391-62
Horizontal
Section 336

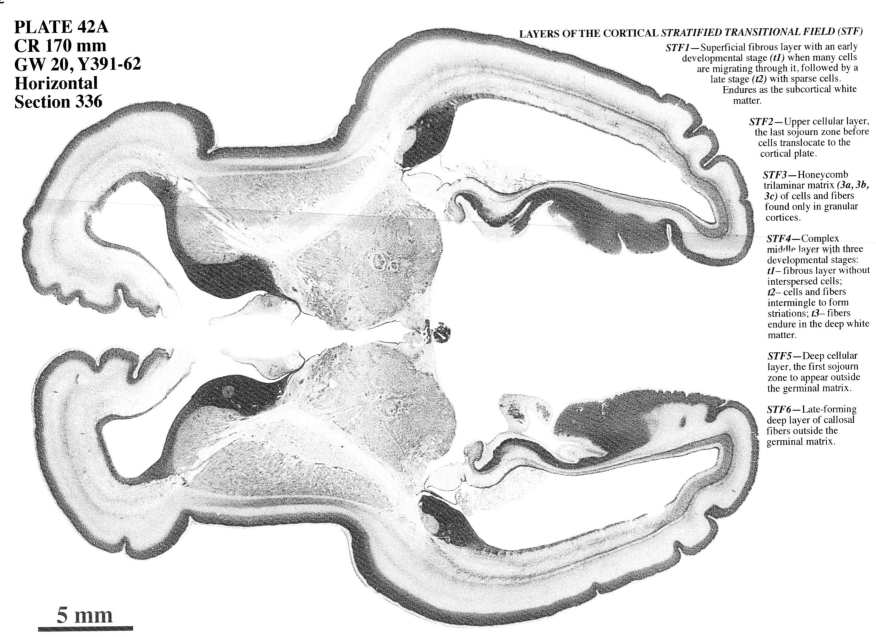

LAYERS OF THE CORTICAL *STRATIFIED TRANSITIONAL FIELD (STF)*

STF1—Superficial fibrous layer with an early developmental stage *(t1)* when many cells are migrating through it, followed by a late stage *(t2)* with sparse cells. Endures as the subcortical white matter.

STF2—Upper cellular layer, the last sojourn zone before cells translocate to the cortical plate.

STF3—Honeycomb trilaminar matrix *(3a, 3b, 3c)* of cells and fibers found only in granular cortices.

STF4—Complex middle layer with three developmental stages: *t1*– fibrous layer without interspersed cells; *t2*– cells and fibers intermingle to form striations; *t3*– fibers endure in the deep white matter.

STF5—Deep cellular layer, the first sojourn zone to appear outside the germinal matrix.

STF6—Late-forming deep layer of callosal fibers outside the germinal matrix.

5 mm

See detail of the brain core in Plates 49A and B.

PLATE 42B

G/EP - glioepithelium/ependyma
GEP - glioepithelium
NEP - neuroepithelium
SVZ - subventricular zone
Germinal and transitional structures in *italics*

Dashed lines at cortical surface are
fixation shrinkage artifacts.

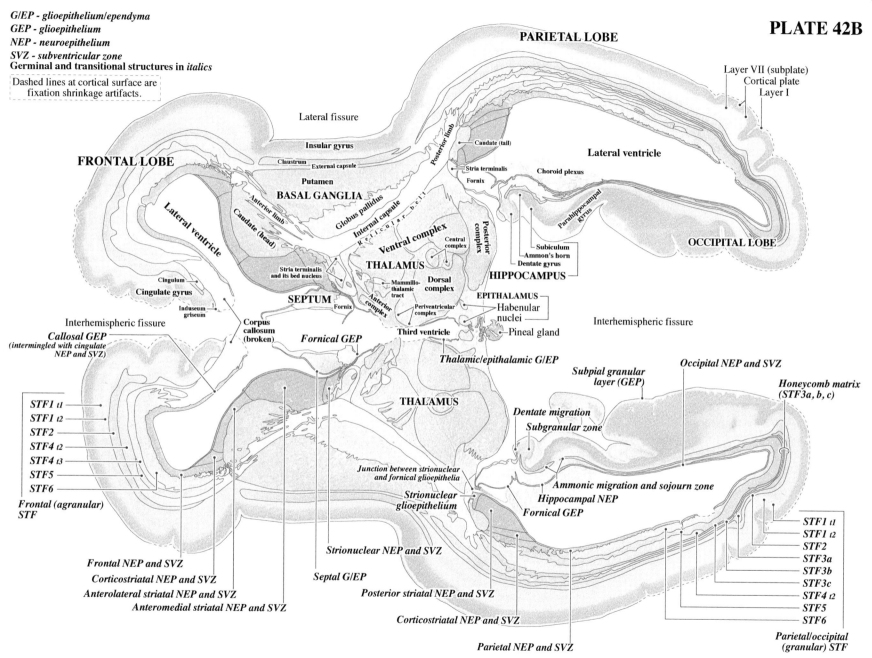

PARIETAL LOBE

Layer VII (subplate)
Cortical plate
Layer I

Lateral fissure

Insular gyrus

Caudate (tail)

Lateral ventricle

FRONTAL LOBE

Claustrum
External capsule

Stria terminalis

Putamen

Fornix

Choroid plexus

BASAL GANGLIA

Globus pallidus

Anterior limb

Lateral ventricle

Caudate (head)

Internal capsule

Reticular belt

Parahippocampal gyrus

OCCIPITAL LOBE

Posterior complex

Central complex

Ventral complex

Subiculum
Ammon's horn
Dentate gyrus

Stria terminalis
and its bed nucleus

THALAMUS

HIPPOCAMPUS

Cingulum

Mammillo-thalamic tract

Dorsal complex

Cingulate gyrus

EPITHALAMUS

Anterior complex

Periventricular complex

Habenular nuclei

Induseum griseum

SEPTUM

Fornix

Pineal gland

Interhemispheric fissure

Corpus callosum (broken)

Third ventricle

Interhemispheric fissure

Callosal GEP
(intermingled with cingulate NEP and SVZ)

Fornical GEP

Third ventricle

Thalamic/epithalamic G/EP

Occipital NEP and SVZ

Subpial granular layer (GEP)

Honeycomb matrix (STF3a, b, c)

THALAMUS

STF1 t1
STF1 t2
STF2
STF4 t2
STF4 t3
STF5
STF6

Dentate migration

Subgranular zone

Junction between strionuclear and fornical glioepithelia

Ammonic migration and sojourn zone

Hippocampal NEP

Fornical GEP

Frontal (agranular) STF

Strionuclear glioepithelium

STF1 t1
STF1 t2
STF2
STF3a
STF3b
STF3c
STF4 t2
STF5
STF6

Frontal NEP and SVZ

Strionuclear NEP and SVZ

Corticostriatal NEP and SVZ

Anterolateral striatal NEP and SVZ

Septal G/EP

Anteromedial striatal NEP and SVZ

Posterior striatal NEP and SVZ

Corticostriatal NEP and SVZ

Parietal/occipital (granular) STF

Parietal NEP and SVZ

PLATE 43A
CR 170 mm
GW 20, Y391-62
Horizontal
Section 386

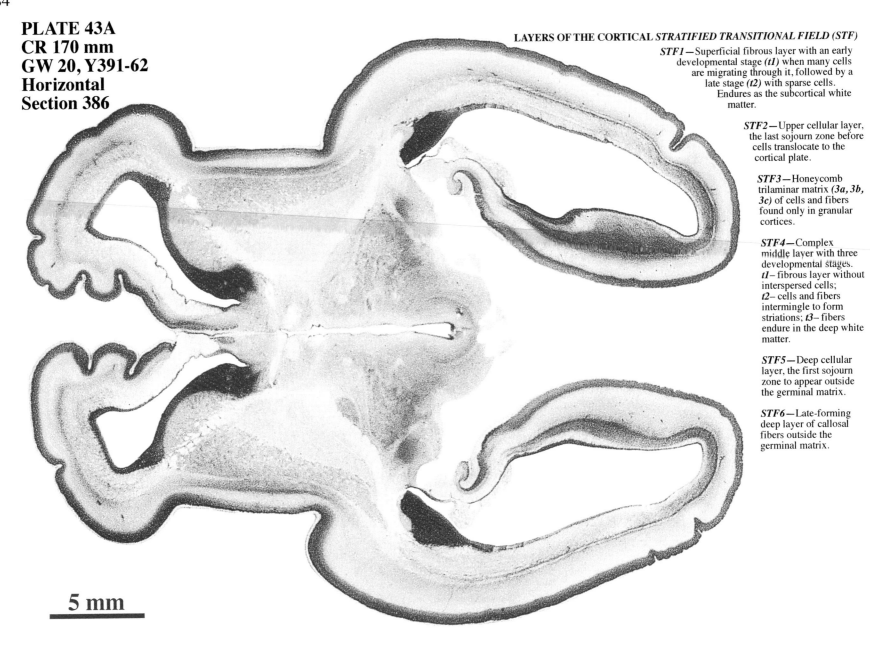

5 mm

LAYERS OF THE CORTICAL *STRATIFIED TRANSITIONAL FIELD (STF)*

STF1—Superficial fibrous layer with an early developmental stage *(t1)* when many cells are migrating through it, followed by a late stage *(t2)* with sparse cells. Endures as the subcortical white matter.

STF2—Upper cellular layer, the last sojourn zone before cells translocate to the cortical plate.

STF3—Honeycomb trilaminar matrix *(3a, 3b, 3c)* of cells and fibers found only in granular cortices.

STF4—Complex middle layer with three developmental stages. *t1*–fibrous layer without interspersed cells; *t2*–cells and fibers intermingle to form striations; *t3*–fibers endure in the deep white matter.

STF5—Deep cellular layer, the first sojourn zone to appear outside the germinal matrix.

STF6—Late-forming deep layer of callosal fibers outside the germinal matrix.

See detail of the brain core in Plates 50A and B.

PLATE 43B

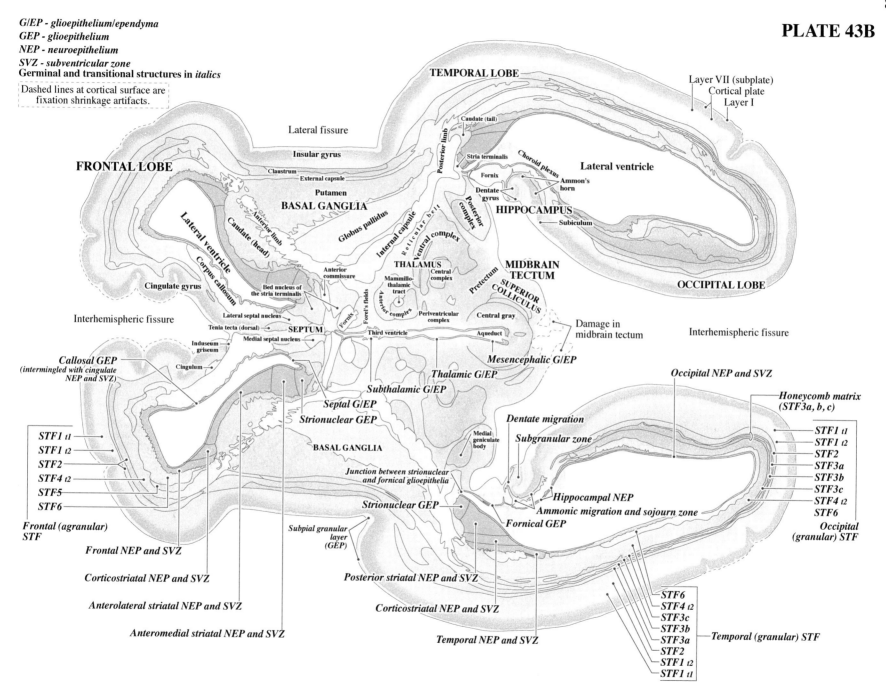

G/EP - glioepithelium/ependyma
GEP - glioepithelium
NEP - neuroepithelium
SVZ - subventricular zone
Germinal and transitional structures in *italics*

Dashed lines at cortical surface are
fixation shrinkage artifacts.

PLATE 44A
CR 170 mm
GW 20, Y391-62
Horizontal
Section 426

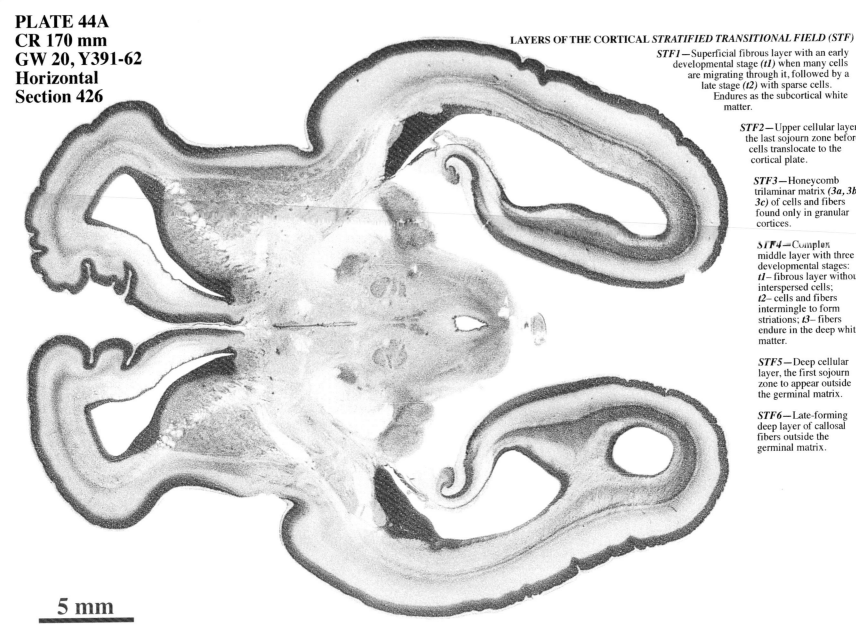

LAYERS OF THE CORTICAL *STRATIFIED TRANSITIONAL FIELD (STF)*

STF1—Superficial fibrous layer with an early developmental stage *(t1)* when many cells are migrating through it, followed by a late stage *(t2)* with sparse cells. Endures as the subcortical white matter.

STF2—Upper cellular layer, the last sojourn zone before cells translocate to the cortical plate.

STF3—Honeycomb trilaminar matrix *(3a, 3b, 3c)* of cells and fibers found only in granular cortices.

STF4—Complex middle layer with three developmental stages: *t1*– fibrous layer without interspersed cells; *t2*– cells and fibers intermingle to form striations; *t3*– fibers endure in the deep white matter.

STF5—Deep cellular layer, the first sojourn zone to appear outside the germinal matrix.

STF6—Late-forming deep layer of callosal fibers outside the germinal matrix.

5 mm

See detail of the brain core in Plates 51A and B.

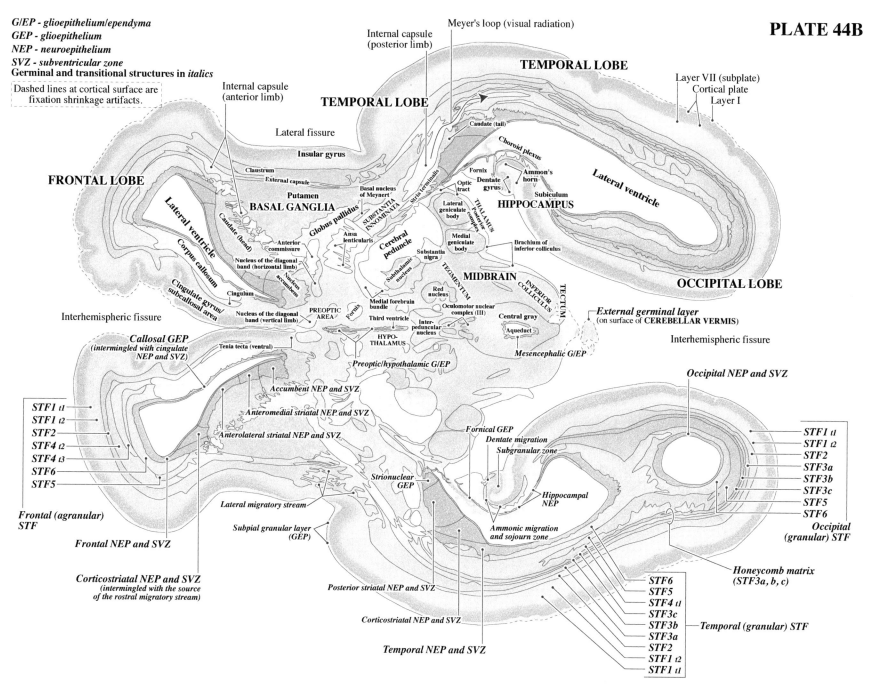

G/EP - *glioepithelium/ependyma*
GEP - *glioepithelium*
NEP - *neuroepithelium*
SVZ - *subventricular zone*
Germinal and transitional structures in *italics*

Dashed lines at cortical surface are
fixation shrinkage artifacts.

Meyer's loop (visual radiation)

Internal capsule
(posterior limb)

TEMPORAL LOBE

PLATE 44B

Layer VII (subplate)
Cortical plate
Layer I

Internal capsule
(anterior limb)

TEMPORAL LOBE

Candate (tail)

Lateral fissure

Insular gyrus

Choroid plexus

Lateral ventricle

Claustrum

External capsule

FRONTAL LOBE

Basal nucleus
of Meynert

Fornix

Dentate
gyrus

Ammon's
horn

Stria terminalis

HIPPOCAMPUS

Subiculum

Lateral ventricle

Putamen

BASAL GANGLIA

Globus pallidus

SUBSTANTIA
INNOMINATA

Optic
tract

Lateral
geniculate
body

THALAMUS
Posterior
complex

Corpus callosum

Candate (head)

Ansa
lenticularis

Cerebral
peduncle

Medial
geniculate
body

Brachium of
inferior colliculus

Anterior
commissure

Substantia
nigra

OCCIPITAL LOBE

*Cingulate gyrus/
subcallosal area*

Nucleus of the diagonal
band (horizontal limb)

Nucleus
accumbens

Subthalamic
nucleus

TEGMENTUM

MIDBRAIN

INFERIOR
COLLICULUS

TECTUM

Cingulum

Red
nucleus

Interhemispheric fissure

Nucleus of the diagonal
band (vertical limb)

PREOPTIC
AREA

Fornix

Medial forebrain
bundle

Oculomotor nuclear
complex (III)

Central gray

External germinal layer
(on surface of **CEREBELLAR VERMIS**)

Callosal GEP
(intermingled with cingulate
NEP and SVZ)

Tenia tecta (ventral)

Third ventricle

HYPO-
THALAMUS

Inter-
peduncular
nucleus

Aqueduct

Mesencephalic G/EP

Interhemispheric fissure

Preoptic/hypothalamic G/EP

Occipital NEP and SVZ

Accumbent NEP and SVZ

STF1 t1
STF1 t2
STF2
STF4 t2
STF4 t3
STF6
STF5

Anteromedial striatal NEP and SVZ

Anterolateral striatal NEP and SVZ

Fornical GEP
Dentate migration
Subgranular zone

STF1 t1
STF1 t2
STF2
STF3a
STF3b
STF3c
STF5
STF6

*Occipital
(granular) STF*

*Strionuclear
GEP*

*Hippocampal
NEP*

*Frontal (agranular)
STF*

Lateral migratory stream

Frontal NEP and SVZ

*Subpial granular layer
(GEP)*

*Ammonic migration
and sojourn zone*

*Honeycomb matrix
(STF3a, b, c)*

Corticostriatal NEP and SVZ
*(intermingled with the source
of the rostral migratory stream)*

Posterior striatal NEP and SVZ

STF6
STF5
STF4 t1
STF3c
STF3b
STF3a
STF2
STF1 t2
STF1 t1

Temporal (granular) STF

Corticostriatal NEP and SVZ

Temporal NEP and SVZ

PLATE 45A
CR 170 mm
GW 20, Y391-62
Horizontal
Section 466

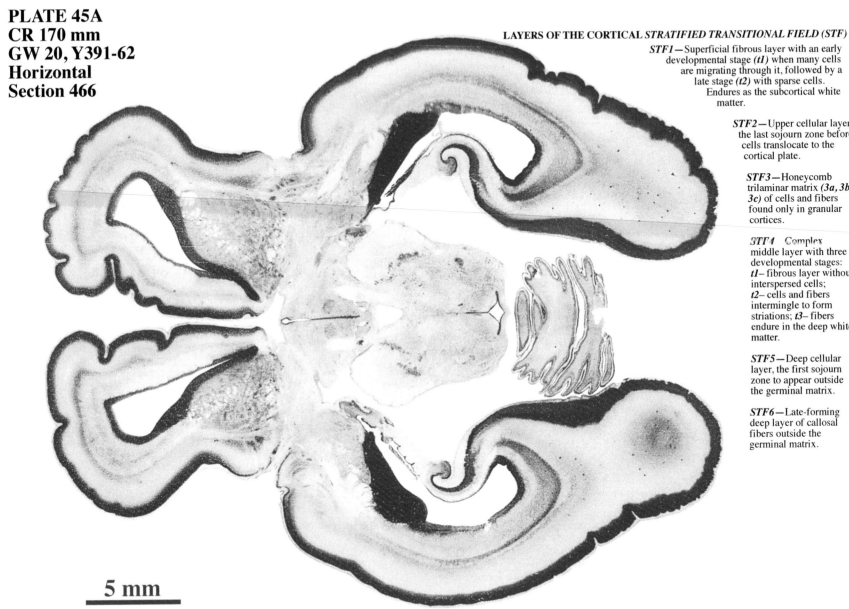

LAYERS OF THE CORTICAL *STRATIFIED TRANSITIONAL FIELD (STF)*

STF1—Superficial fibrous layer with an early developmental stage *(t1)* when many cells are migrating through it, followed by a late stage *(t2)* with sparse cells. Endures as the subcortical white matter.

STF2—Upper cellular layer, the last sojourn zone before cells translocate to the cortical plate.

STF3—Honeycomb trilaminar matrix *(3a, 3b, 3c)* of cells and fibers found only in granular cortices.

STF4—Complex middle layer with three developmental stages: *t1*– fibrous layer without interspersed cells; *t2*– cells and fibers intermingle to form striations; *t3*– fibers endure in the deep white matter.

STF5—Deep cellular layer, the first sojourn zone to appear outside the germinal matrix.

STF6—Late-forming deep layer of callosal fibers outside the germinal matrix.

5 mm

See detail of the brain core in Plates 52A and B.

G/EP - glioepithelium/ependyma
GEP - glioepithelium
NEP - neuroepithelium
SVZ - subventricular zone
Germinal and transitional structures in *italics*

Dashed lines at cortical surface are
fixation shrinkage artifacts.

TEMPORAL LOBE

Honeycomb matrix
(STF3a, b, c)

TEMPORAL LOBE

Internal capsule
(posterior limb)

Choroid
plexus

Lateral fissure

Ammon's
horn

Lateral
ventricle

Internal capsule
(anterior limb)

Insular gyrus

Endopiriform
nucleus

Fornix

Dentate
gyrus

Subiculum

Layer I

Claustrum
External capsule

HIPPOCAMPUS

Cortical plate

FRONTAL LOBE

**Lateral
ventricle**

Putamen

Stria
terminalis

Layer VII (subplate)

BASAL GANGLIA

AMYGDALA

OCCIPITAL LOBE

Caudate (tail)

Corpus callosum

Caudate (head)

SUBSTANTIA
INNOMINATA

Anterior commissure

TEGMENTUM

MIDBRAIN

Cerebral
peduncle

Lateral
olfactory
tract

Optic
tract

Substantia
nigra

Lateral lemniscus

TECTUM

Nucleus
accumbens

**INFERIOR
COLLICULUS**

External germinal layer

Tenia tecta (ventral)

Fornix

Superior
cerebellar
peduncle

Central gray

Subcallosal area

PREOPTIC
AREA

**HYPO-
THALAMUS**

Medial
forebrain
bundle

Inter-
peduncular
nucleus

Aqueduct

**CEREBELLUM
(VERMIS)**

Interhemispheric fissure

Interhemispheric fissure

Third ventricle

Preoptic/hypothalamic G/EP

Accumbent NEP and SVZ

Anterolateral striatal NEP and SVZ

Perifascicular GEP
(surrounds fiber tracts)

Callosal GEP

*Mesencephalic
G/EP*

STF6

*Lateral
migratory
stream*
*(invades primary
olfactory cortex)*

STF1 t1
STF1 t2
STF2
STF3a

STF5

STF4 t2

Fornical GEP

STF4 t1

Dentate migration

*Posterior striatal
NEP and SVZ*

VENTRAL
HIPPOCAMPUS

STF2

Subgranular zone

*Occipital
(granular) STF*

STF1 t2

Ammonic migration and sojourn zone

STF1 t1

Hippocampal NEP

**Orbitofrontal
(agranular) STF**

*Primary olfactory
cortex (piriform)*

STF1 t1
STF1 t2

Orbitofrontal NEP and SVZ

Corticostriatal NEP and SVZ
*(intermingled with the source
of the rostral migratory stream)*

STF2

STF3a

Subpial granular layer
(GEP)

STF3b

Temporal (granular) STF

Corticostriatal NEP and SVZ

STF3c

STF4 t2

STF5

STF6

Temporal NEP and SVZ

PLATE 46A
CR 170 mm
GW 20, Y391-62
Horizontal
Section 506

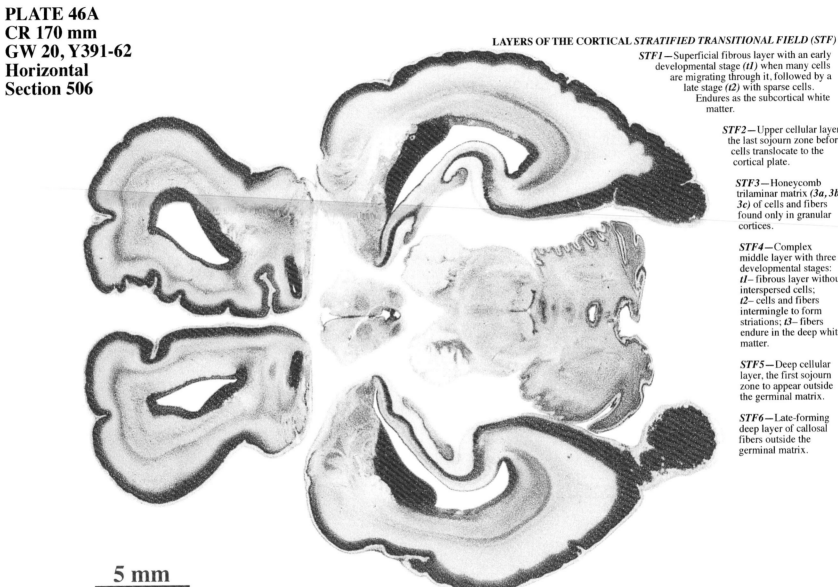

5 mm

LAYERS OF THE CORTICAL *STRATIFIED TRANSITIONAL FIELD (STF)*

STF1—Superficial fibrous layer with an early developmental stage *(t1)* when many cells are migrating through it, followed by a late stage *(t2)* with sparse cells. Endures as the subcortical white matter.

STF2—Upper cellular layer, the last sojourn zone before cells translocate to the cortical plate.

STF3—Honeycomb trilaminar matrix *(3a, 3b, 3c)* of cells and fibers found only in granular cortices.

STF4—Complex middle layer with three developmental stages: *t1*– fibrous layer without interspersed cells; *t2*– cells and fibers intermingle to form striations; *t3*– fibers endure in the deep white matter.

STF5—Deep cellular layer, the first sojourn zone to appear outside the germinal matrix.

STF6—Late-forming deep layer of callosal fibers outside the germinal matrix.

See detail of the brain core in Plates 53A and B.

PLATE 46B

G/EP - glioepithelium/ependyma
GEP - glioepithelium
NEP - neuroepithelium
SVZ - subventricular zone
Germinal and transitional structures in *italics*

Dashed lines at cortical surface are
fixation shrinkage artifacts.

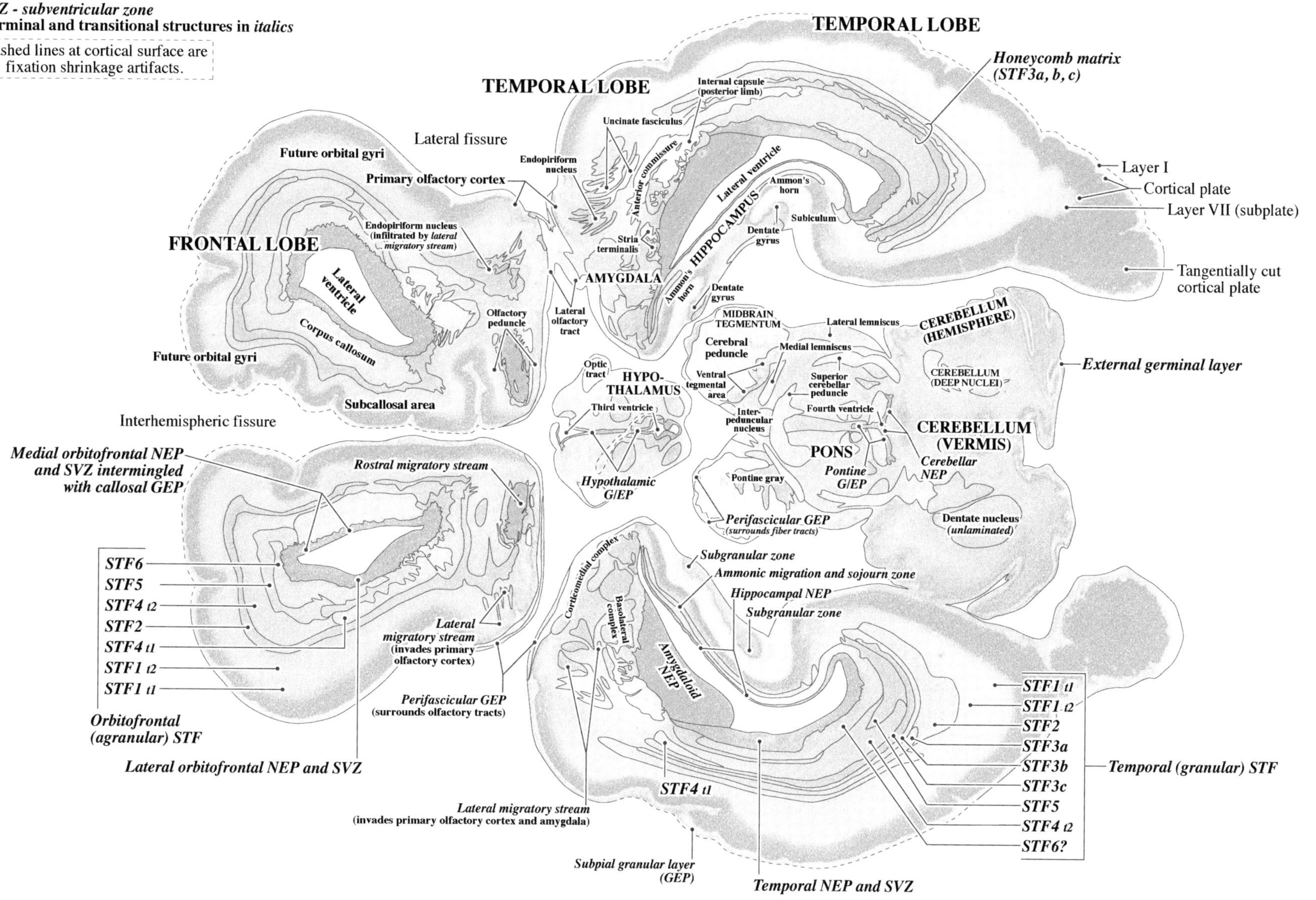

TEMPORAL LOBE

Honeycomb matrix
(STF3a, b, c)

TEMPORAL LOBE

Internal capsule
(posterior limb)

Uncinate fasciculus

Lateral fissure

Future orbital gyri

Endopiriform
nucleus

Primary olfactory cortex

Lateral ventricle

Ammon's
horn

Layer I

Cortical plate

Layer VII (subplate)

Endopiriform nucleus
(infiltrated by *lateral*
migratory stream)

FRONTAL LOBE

Lateral
ventricle

Stria
terminalis

AMYGDALA

Ammon's
horn

HIPPOCAMPUS

Subiculum

Dentate
gyrus

Tangentially cut
cortical plate

Corpus callosum

Olfactory
peduncle

Lateral
olfactory
tract

Dentate
gyrus

**MIDBRAIN
TEGMENTUM**

Lateral lemniscus

**CEREBELLUM
(HEMISPHERE)**

Cerebral
peduncle

Medial lemniscus

Future orbital gyri

Ammon's
horn

Optic
tract

**HYPO-
THALAMUS**

Ventral
tegmental
area

Superior
cerebellar
peduncle

**CEREBELLUM
(DEEP NUCLEI)**

External germinal layer

Subcallosal area

Third ventricle

Inter-
peduncular
nucleus

Fourth ventricle

**CEREBELLUM
(VERMIS)**

Interhemispheric fissure

*Hypothalamic
G/EP*

PONS

Pontine gray

*Pontine
G/EP*

*Cerebellar
NEP*

*Medial orbitofrontal NEP
and SVZ intermingled
with callosal GEP*

Rostral migratory stream

Perifascicular GEP
(surrounds fiber tracts)

*Dentate nucleus
(unlaminated)*

STF6

STF5

STF4 t2

STF2

STF4 t1

STF1 t2

STF1 t1

*Lateral
migratory stream*
(invades primary
olfactory cortex)

Subgranular zone

Ammonic migration and sojourn zone

Hippocampal NEP

Subgranular zone

Cortico-
medial
complex

Baso-
lateral
complex

*Amygdaloid
NEP*

STF1 t1

STF1 t2

STF2

STF3a

STF3b

STF3c

STF5

STF4 t2

STF6?

*Orbitofrontal
(agranular) STF*

Perifascicular GEP
(surrounds olfactory tracts)

Lateral orbitofrontal NEP and SVZ

Temporal (granular) STF

STF4 t1

Lateral migratory stream
(invades primary olfactory cortex and amygdala)

Subpial granular layer
(GEP)

Temporal NEP and SVZ

92

**PLATE 47A
CR 170 mm
GW 20, Y391-62
Horizontal
Section 546**

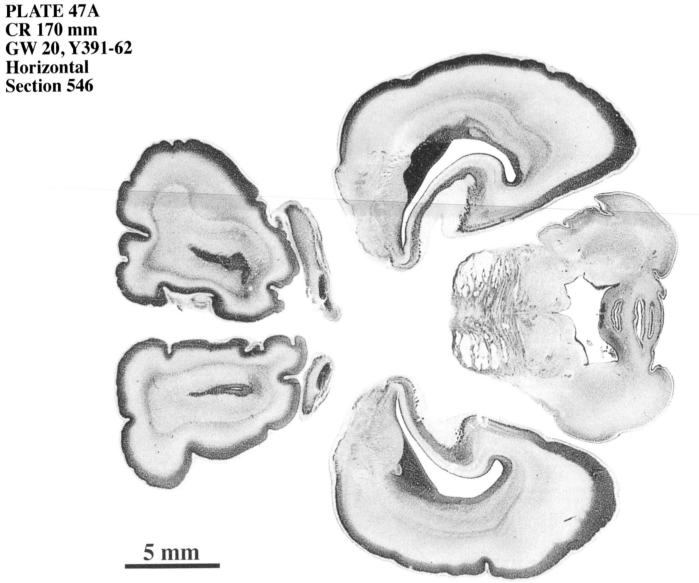

5 mm

**LAYERS OF THE CORTICAL
*STRATIFIED TRANSITIONAL
FIELD (STF)***

STF1—Superficial fibrous
layer with an early
developmental stage *(t1)*
when many cells are
migrating through it, followed
by a late stage *(t2)* with sparse
cells. Endures as the
subcortical white matter.

STF2—Upper cellular layer,
the last sojourn zone before
cells translocate to the cortical
plate.

STF3—Honeycomb
trilaminar matrix *(3a, 3b, 3c)*
of cells and fibers found only
in granular cortices.

STF4—Complex middle
layer with three
developmental stages:
t1– fibrous layer without
interspersed cells;
t2– cells and fibers
intermingle to form striations;
t3– fibers endure in the deep
white matter.

STF5—Deep cellular layer,
the first sojourn zone to
appear outside the germinal
matrix.

STF6—Late-forming deep
layer of callosal fibers outside
the germinal matrix.

See detail of the brain core in Plates 54A and B.

PLATE 47B

G/EP - glioepithelium/ependyma
GEP - glioepithelium
NEP - neuroepithelium
SVZ - subventricular zone
Germinal and transitional structures in *italics*

Dashed lines at cortical surface are
fixation shrinkage artifacts.

TEMPORAL LOBE

Layer I
Cortical plate
Layer VII (subplate)

Lateral fissure

Uncinate fasciculus
Endopiriform nucleus
Anterior commissure
Lateral ventricle
Ammon's horn

Future orbital gyri

Primary olfactory cortex

Subiculum

AMYGDALA
HIPPOCAMPUS

Entorhinal cortex

CEREBELLUM (HEMISPHERE)

FRONTAL LOBE

Subiculum

CEREBELLUM (DEEP NUCLEI)

Medial lemniscus
Lateral lemniscus

Olfactory peduncle
Lateral olfactory tract
Entorhinal cortex

Superior cerebellar peduncle

Future orbital gyri

Transpontine corticofugal tract

Principal sensory nucleus (V)

Subcallosal area

Pontocerebellar fibers

PONS

Fourth ventricle

CEREBELLUM (VERMIS)

Interhemispheric fissure

Pontocerebellar fibers (decussation)

Pontine G/EP

Germinal trigone

Orbitofrontal NEP and SVZ

Rostral migratory stream

Pontine gray

Cerebellar NEP

Anterior extramural migratory stream

Trigeminalmotor nucleus (V)

Middle cerebellar peduncle

STF1 t1
STF1 t2
STF2
STF4 t2
STF5
STF6?

Dentate nucleus

External germinal layer

Ammonic migration and sojourn zone

Corticomedial complex

Hippocampal NEP

Inferior cerebellar peduncle

Orbitofrontal (agranular) STF

Basolateral complex

Temporal (granular) STF

Lateral migratory stream
(invades primary olfactory cortex and amygdala)

STF1 t1
STF1 t2
STF2
STF4 t2
STF5
STF6
STF4 t1
STF3a
STF3b
STF3c

Amygdaloid NEP

Subpial granular layer (GEP)

Temporal NEP and SVZ

Honeycomb matrix

94

PLATE 48A
CR 170 mm
GW 20, Y391-62
Horizontal
Section 596

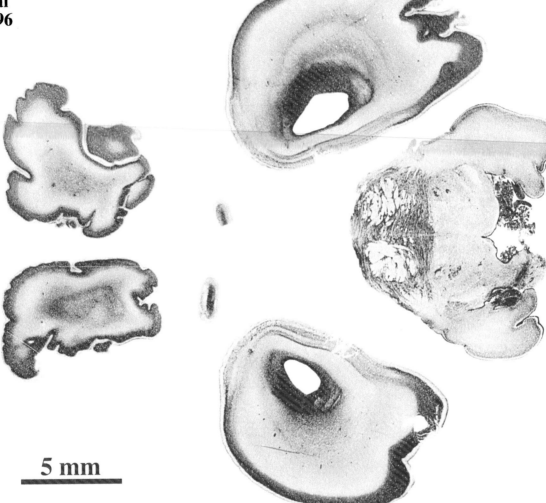

LAYERS OF THE CORTICAL
STRATIFIED TRANSITIONAL
FIELD (STF)

STF1—Superficial fibrous
layer with an early
developmental stage *(t1)*
when many cells are
migrating through it, followed
by a late stage *(t2)* with sparse
cells. Endures as the
subcortical white matter.

STF2—Upper cellular layer,
the last sojourn zone before
cells translocate to the cortical
plate.

STF3—Honeycomb
trilaminar matrix *(3a, 3b, 3c)*
of cells and fibers found only
in granular cortices.

STF4—Complex middle
layer with three
developmental stages:
t1– fibrous layer without
interspersed cells;
t2– cells and fibers
intermingle to form striations;
t3– fibers endure in the deep
white matter.

STF5—Deep cellular layer,
the first sojourn zone to
appear outside the germinal
matrix.

5 mm

See detail of the brain core in Plates 55A and B.

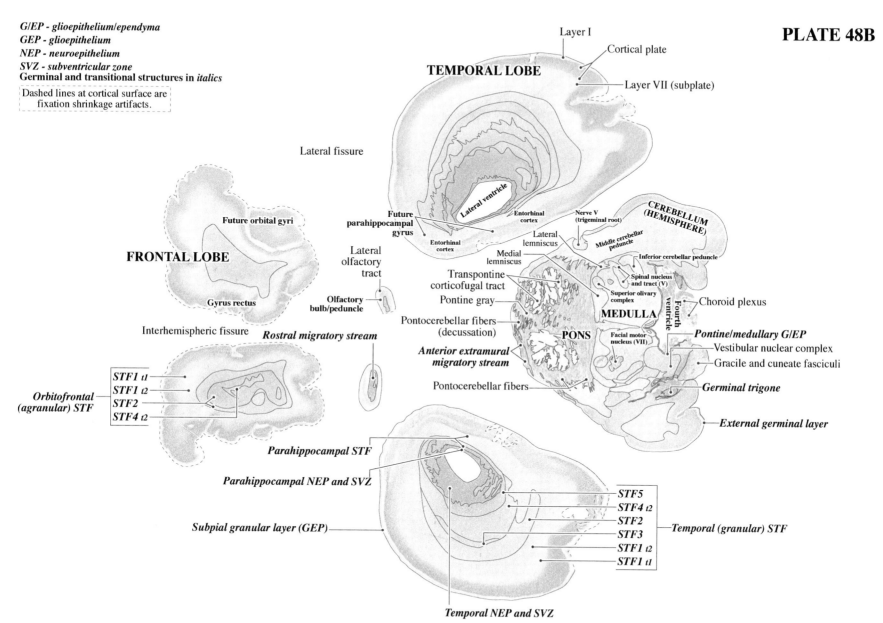

G/EP - *glioepithelium/ependyma*
GEP - *glioepithelium*
NEP - *neuroepithelium*
SVZ - *subventricular zone*
Germinal and transitional structures in *italics*

Dashed lines at cortical surface are
fixation shrinkage artifacts.

Layer I
Cortical plate
Layer VII (subplate)

TEMPORAL LOBE

Lateral fissure

Future orbital gyri

FRONTAL LOBE

Gyrus rectus

Lateral
olfactory
tract

Future
parahippocampal
gyrus

Lateral ventricle

Entorhinal
cortex

Entorhinal
cortex

Nerve V
(trigeminal root)

**CEREBELLUM
(HEMISPHERE)**

Middle cerebellar
peduncle

Inferior cerebellar peduncle

Lateral
lemniscus

Medial
lemniscus

Spinal nucleus
and tract (V)

Transpontine
corticofugal tract

Superior olivary
complex

MEDULLA

Pontine gray

Olfactory
bulb/peduncle

Interhemispheric fissure

Rostral migratory stream

Pontocerebellar fibers
(decussation)

PONS

Facial motor
nucleus (VII)

Fourth
ventricle

Choroid plexus

Pontine/medullary G/EP

Vestibular nuclear complex

Gracile and cuneate fasciculi

STF1 t1
STF1 t2
STF2
STF4 t2

*Orbitofrontal
(agranular) STF*

*Anterior extramural
migratory stream*

Pontocerebellar fibers

Germinal trigone

External germinal layer

Parahippocampal STF

Parahippocampal NEP and SVZ

Subpial granular layer (GEP)

STF5
STF4 t2
STF2
STF3
STF1 t2
STF1 t1

Temporal (granular) STF

Temporal NEP and SVZ

**PLATE 49A
CR 170 mm
GW 20, Y391-62
Horizontal
Section 336**

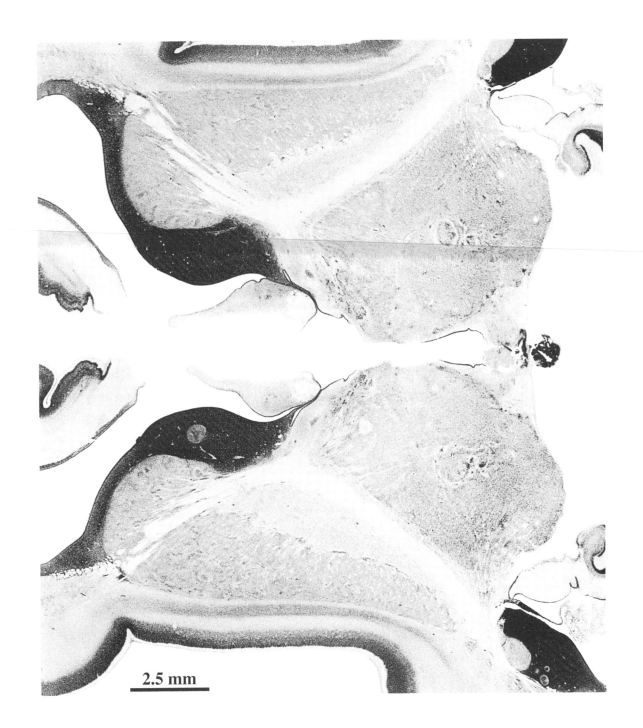

2.5 mm

**See the entire section
in Plates 42A and B.**

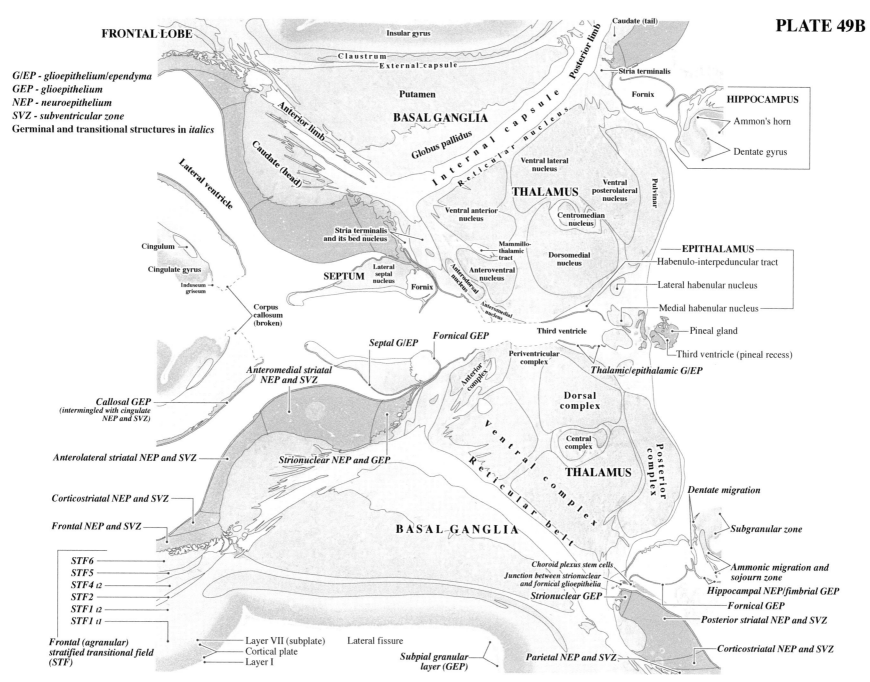

FRONTAL LOBE

Insular gyrus

Claustrum
External capsule

G/EP - *glioepithelium/ependyma*
GEP - *glioepithelium*
NEP - *neuroepithelium*
SVZ - *subventricular zone*
Germinal and transitional structures in *italics*

Putamen

BASAL GANGLIA

Globus pallidus

Anterior limb

Caudate (head)

Cingulum

Cingulate gyrus

Induseum griseum

Corpus callosum (broken)

Lateral ventricle

Stria terminalis and its bed nucleus

SEPTUM

Lateral septal nucleus

Fornix

Caudate (tail)

Stria terminalis

Posterior limb

Internal capsule

Reticular nucleus

Fornix

HIPPOCAMPUS

Ammon's horn

Dentate gyrus

Ventral lateral nucleus

THALAMUS

Ventral posterolateral nucleus

Pulvinar

Ventral anterior nucleus

Centromedian nucleus

Mammillo-thalamic tract

Anterodorsal nucleus

Anteroventral nucleus

Dorsomedial nucleus

Anteromedial nucleus

EPITHALAMUS

Habenulo-interpeduncular tract

Lateral habenular nucleus

Medial habenular nucleus

Pineal gland

Third ventricle (pineal recess)

Third ventricle

Septal G/EP

Fornical GEP

Periventricular complex

Thalamic/epithalamic G/EP

Anteromedial striatal NEP and SVZ

Anterior complex

Dorsal complex

Callosal GEP (intermingled with cingulate NEP and SVZ)

Anterolateral striatal NEP and SVZ

Strionuclear NEP and GEP

Central complex

Posterior complex

Reticular belt

Ventral complex

THALAMUS

Corticostriatal NEP and SVZ

Frontal NEP and SVZ

Dentate migration

Subgranular zone

STF6
STF5
STF4 t2
STF2
STF1 t2
STF1 t1

BASAL GANGLIA

Choroid plexus stem cells

Junction between strionuclear and fornical glioepithelia

Strionuclear GEP

Ammonic migration and sojourn zone

Hippocampal NEP/fimbrial GEP

Fornical GEP

Posterior striatal NEP and SVZ

Frontal (agranular) stratified transitional field (STF)

Layer VII (subplate)
Cortical plate
Layer I

Lateral fissure

Subpial granular layer (GEP)

Parietal NEP and SVZ

Corticostriatal NEP and SVZ

PLATE 50A
CR 170 mm
GW 20, Y391-62
Horizontal
Section 386

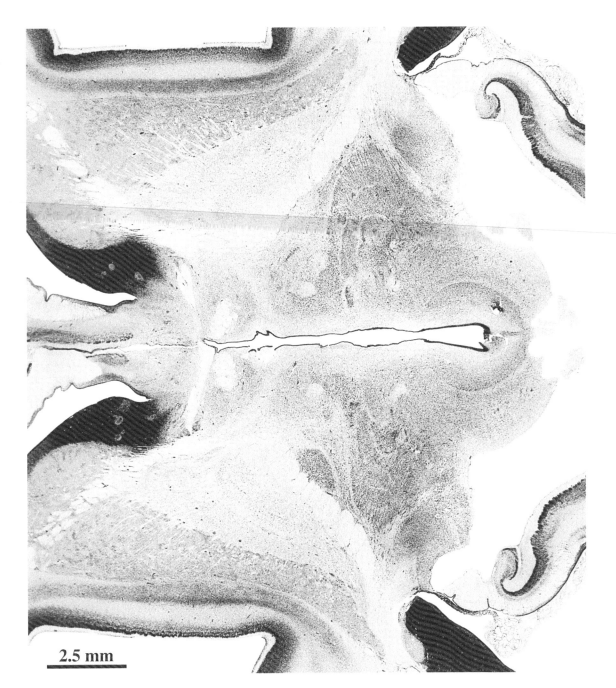

2.5 mm

See the entire section
in Plates 43A and B.

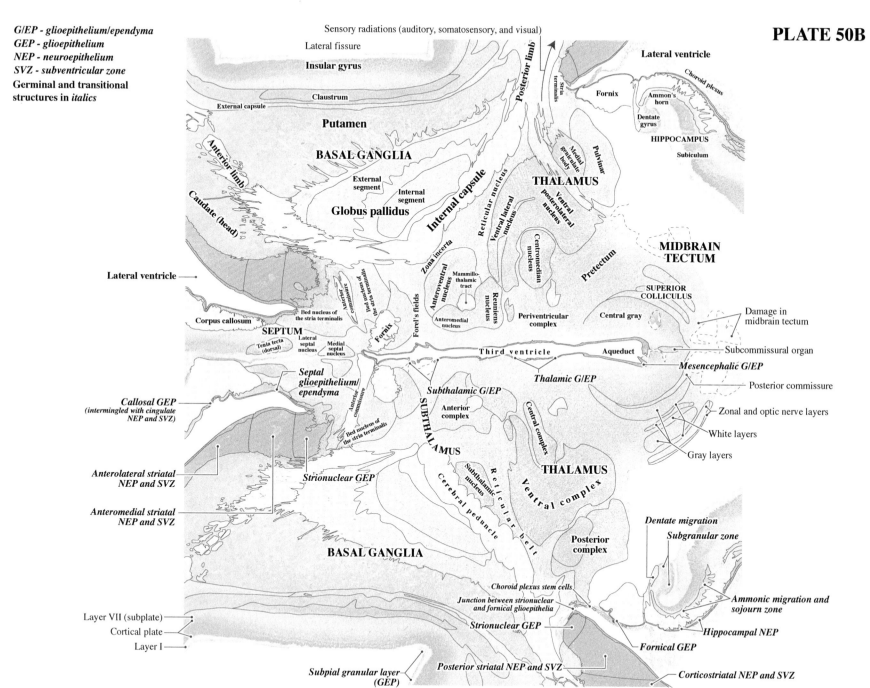

G/EP - glioepithelium/ependyma
GEP - glioepithelium
NEP - neuroepithelium
SVZ - subventricular zone
Germinal and transitional structures in *italics*

Sensory radiations (auditory, somatosensory, and visual)

Lateral fissure

Insular gyrus

Claustrum

External capsule

Putamen

BASAL GANGLIA

Anterior limb

External segment

Internal segment

Caudate (head)

Globus pallidus

Internal capsule

Lateral ventricle

Corpus callosum

Bed nucleus of the stria terminalis

SEPTUM

Tenia tecta (dorsal)

Lateral septal nucleus

Medial septal nucleus

Anterior commissure

Fornix

Forel's fields

Septal glioepithelium/ ependyma

Callosal GEP (intermingled with cingulate NEP and SVZ)

Bed nucleus of the stria terminalis

Anterolateral striatal NEP and SVZ

Anteromedial striatal NEP and SVZ

Strionuclear GEP

Subthalamic G/EP

SUBTHALAMUS

Anterior complex

Subthalamic nucleus

Cerebral peduncle

Reticular belt

BASAL GANGLIA

Layer VII (subplate)

Cortical plate

Layer I

Subpial granular layer (GEP)

Posterior limb

Stria terminalis

Fornix

Lateral ventricle

Choroid plexus

Ammon's horn

Dentate gyrus

HIPPOCAMPUS

Subiculum

Medial geniculate body

Pulvinar

THALAMUS

Reticular nucleus

Ventral lateral nucleus

Ventral posterolateral nucleus

Centromedian nucleus

Zona incerta

Mammillo-thalamic tract

Anteroventral nucleus

Reuniens nucleus

Anteromedial nucleus

Periventricular complex

Pretectum

Central gray

MIDBRAIN TECTUM

SUPERIOR COLLICULUS

Damage in midbrain tectum

Third ventricle

Aqueduct

Thalamic G/EP

Mesencephalic G/EP

Subcommissural organ

Posterior commissure

Zonal and optic nerve layers

White layers

Gray layers

Central complex

THALAMUS

Ventral complex

Posterior complex

Dentate migration

Subgranular zone

Choroid plexus stem cells

Junction between strionuclear and fornical glioepithelia

Strionuclear GEP

Ammonic migration and sojourn zone

Hippocampal NEP

Fornical GEP

Posterior striatal NEP and SVZ

Corticostriatal NEP and SVZ

PLATE 51A
CR 170 mm, GW 20, Y391-62, Horizontal, Section 426

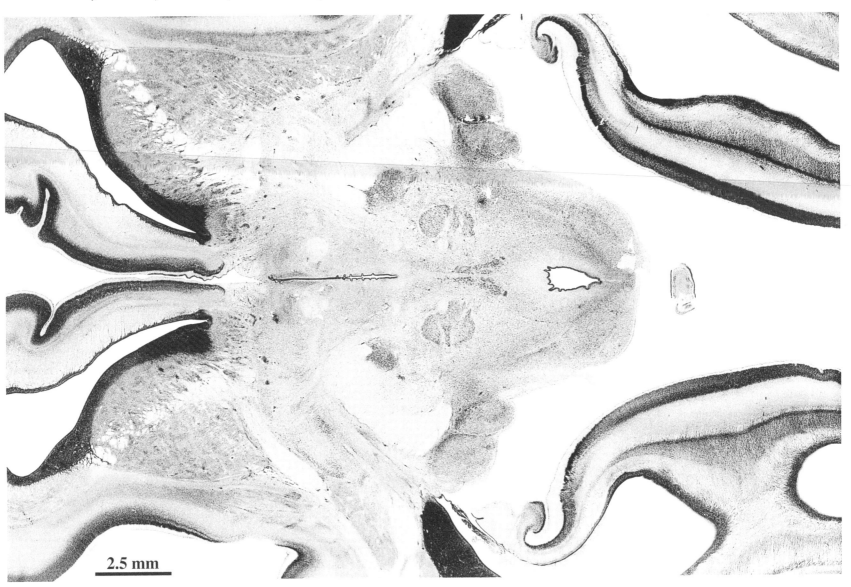

2.5 mm

See the entire section in Plates 44A and B.

G/EP - glioepithelium/ependyma
GEP - glioepithelium
NEP - neuroepithelium
SVZ - subventricular zone

Germinal and transitional
structures in *italics*

Meyer's loop (visual radiation)
Internal capsule (anterior limb)
Internal capsule (posterior limb)

FRONTAL LOBE

Claustrum
External capsule
Putamen

Basal nucleus
of Meynert
SUBSTANTIA
INNOMINATA

Ammon's
horn
Fornix
Dentate
gyrus
HIPPOCAMPUS
Subiculum

TEMPORAL LOBE

Lateral ventricle

Optic tract
Optic tract
Optic tract
Lateral
geniculate
body

BASAL GANGLIA
Globus pallidus
Ansa lenticularis
Cerebral peduncle

Medial
geniculate
body

THALAMUS

Lateral ventricle
Caudate (head)

Anterior commissure
Nucleus of the diagonal band (horizontal limb)
Forel's fields

Substantia nigra
Subthalamic nucleus

Superior cerebellar peduncle
Parabrachial nucleus

Brachium of
inferior colliculus

OCCIPITAL LOBE

Layer VII (subplate)
Cortical plate
Layer I

Corpus callosum
Cingulate gyrus/subcallosal area
Cingulum
Nucleus accumbens

Medial forebrain bundle
Lateral preoptic and hypothalamic areas

Red nucleus

Mesencephalic nucleus (V)
Central nucleus

Damage in midbrain tectum
External germinal layer
(on surface of CEREBELLAR VERMIS)

Nucleus of the diagonal band (vertical limb)
Medial preoptic nucleus
Fornix
Paraventricular nucleus
Third ventricle

Habenulo-interpeduncular tract
Inter-peduncular nucleus
Oculomotor nuclear complex (III)
Central gray
Aqueduct

Culmen (IV-V)

Callosal GEP (intermingled with cingulate NEP and SVZ)
Tenia tecta (ventral)

PREOPTIC AREA
Preoptic/hypothalamic G/EP
HYPO-THALAMUS

Medial longitudinal fasciculus
Mesencephalic G/EP

INFERIOR COLLICULUS

Occipital (granular) stratified transitional field (STF)

MIDBRAIN
TECTUM

STF1 t1
STF1 t2
STF2?
STF3a
STF3b
STF3c
STF5
STF6

Accumbent NEP and SVZ
Anteromedial striatal NEP and SVZ

TEGMENTUM
Cerebral peduncle

Frontal (agranular) stratified transitional field (STF)
STF6
STF5?
STF4 t2
STF1 t2
STF1 t1

Anterolateral striatal NEP and SVZ
Corticostriatal NEP and SVZ (intermingled with the source of the rostral migratory stream)
Frontal NEP and SVZ

Basal nucleus of Meynert
Stria terminalis
Optic tract

THALAMUS Posterior complex
Junction between strionuclear and perifascicular glioepithelia
Choroid plexus stem cells
Fornical GEP
Dentate migration
Subgranular zone

Occipital NEP and SVZ

STF2 (blends with claustrum and may contribute cells to the lateral migratory stream)
Lateral fissure
Insular gyrus
Lateral migratory stream
Subpial granular layer (GEP) →

Perifascicular GEP (invades fiber tracts)
Internal capsule
Strionuclear GEP
Caudate (tail)

Hippocampal NEP
Ammonic migration and sojourn zone
Temporal NEP and SVZ

STF6
STF5
STF4 t1
STF3c
STF3b
STF3a

Posterior striatal NEP and SVZ
Corticostriatal NEP and SVZ

Temporal (granular) stratified transitional field (STF)

PLATE 52A
CR 170 mm, GW 20, Y391-62, Horizontal, Section 466

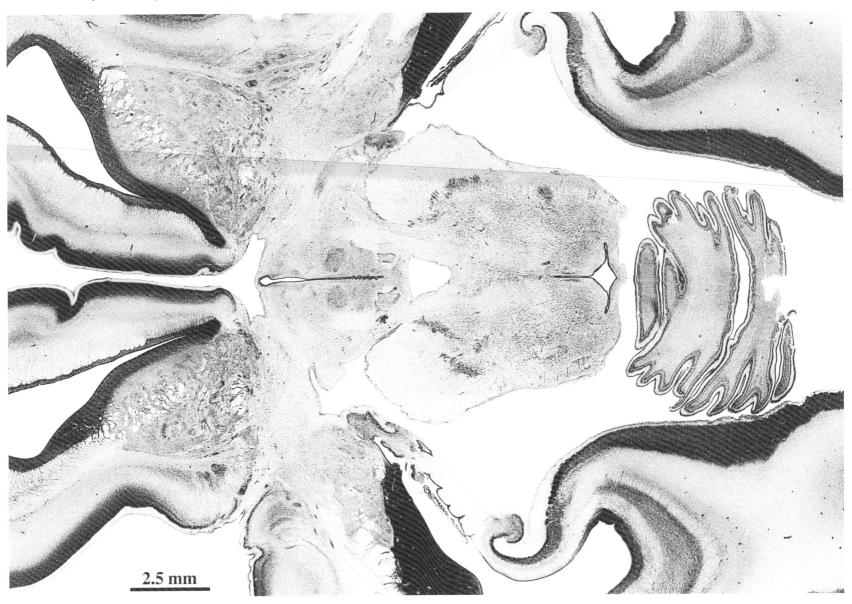

2.5 mm

See the entire section in Plates 45A and B.

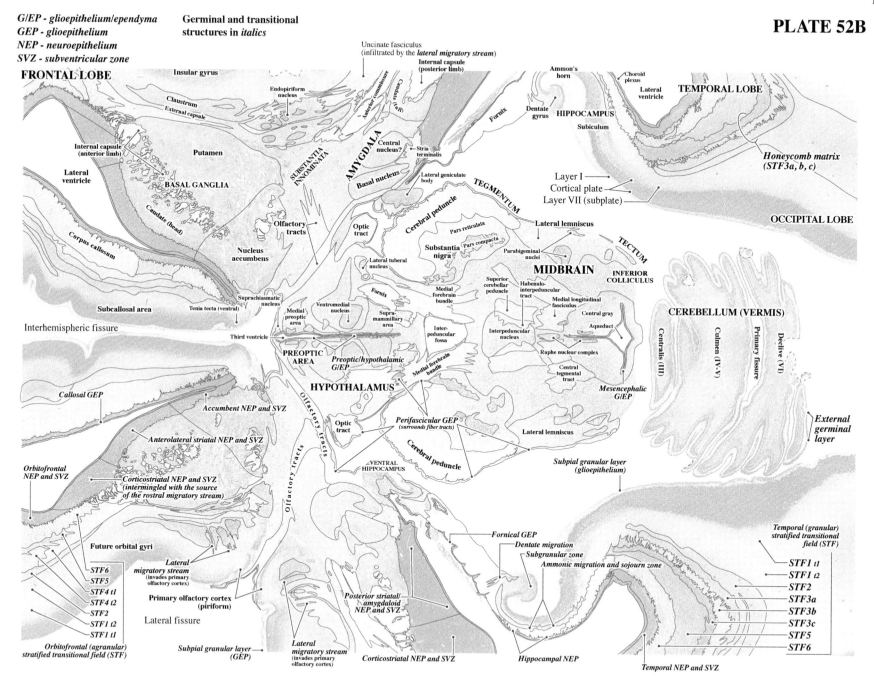

G/EP - glioepithelium/ependyma
GEP - glioepithelium
NEP - neuroepithelium
SVZ - subventricular zone

Germinal and transitional
structures in *italics*

104

CR 170 mm, GW 20, Y391-62, Horizontal, Section 506

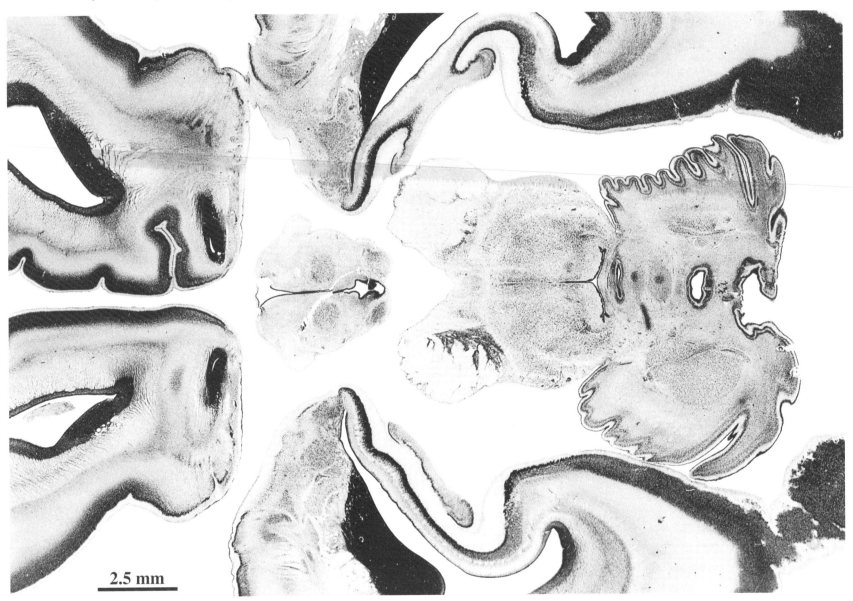

2.5 mm

See the entire section in Plates 46A and B.

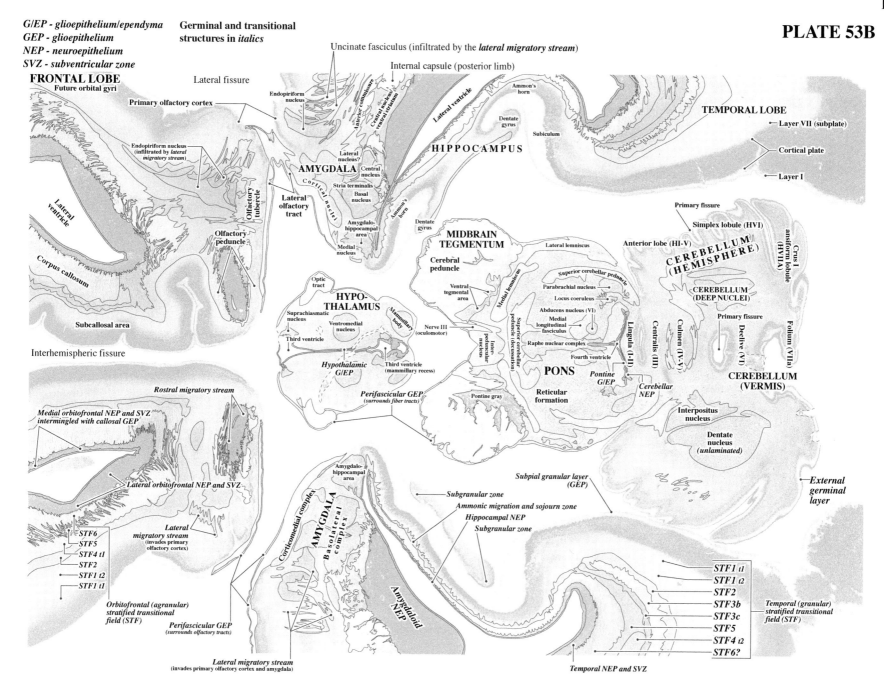

G/EP - glioepithelium/ependyma
GEP - glioepithelium
NEP - neuroepithelium
SVZ - subventricular zone

Germinal and transitional
structures in *italics*

PLATE 54A
CR 170 mm, GW 20
Y391-62
Horizontal
Section 546

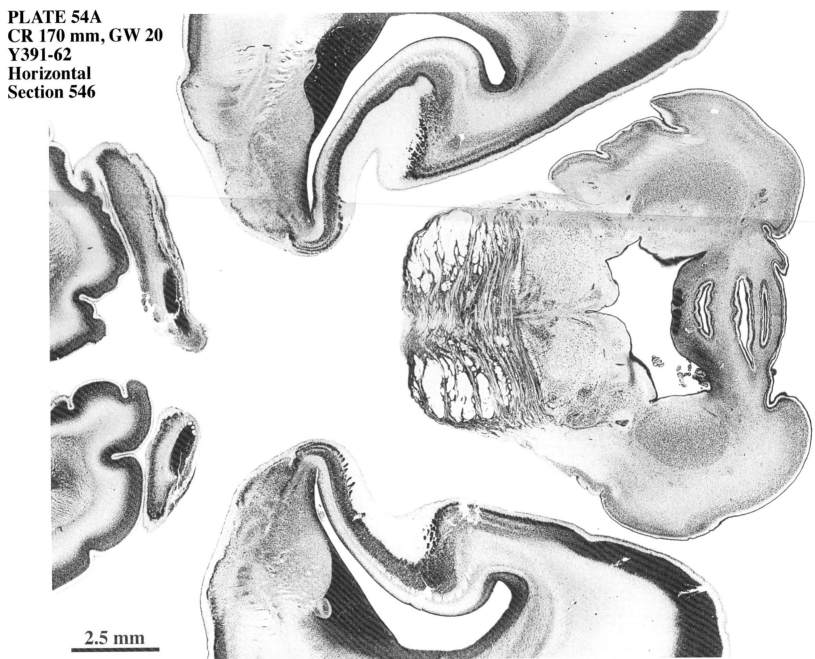

2.5 mm

See the entire section in Plates 47A and B.

PLATE 54B

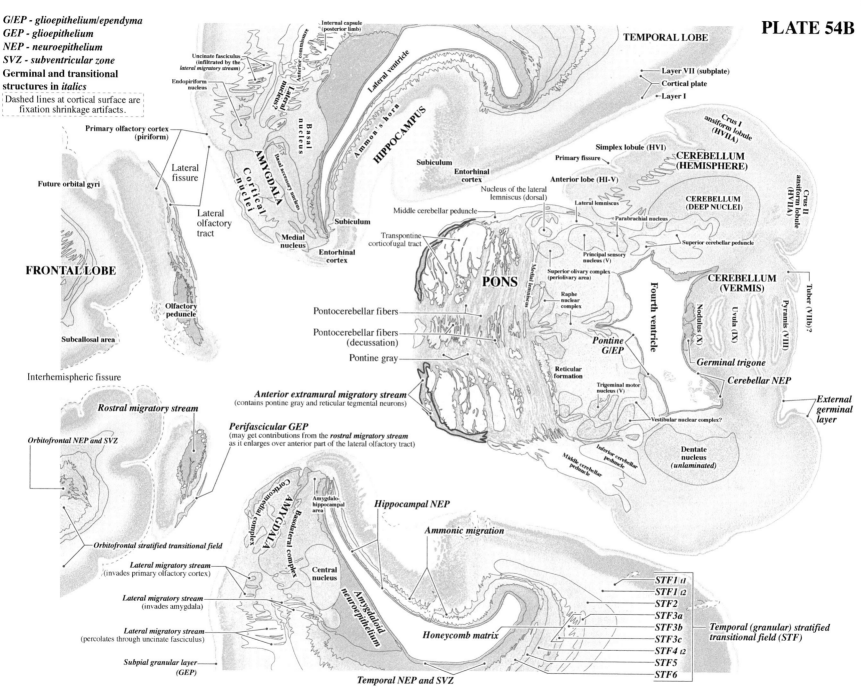

G/EP - *glioepithelium/ependyma*
GEP - *glioepithelium*
NEP - *neuroepithelium*
SVZ - *subventricular zone*
Germinal and transitional structures in *italics*

Dashed lines at cortical surface are fixation shrinkage artifacts.

Internal capsule (posterior limb)

TEMPORAL LOBE

Uncinate fasciculus (infiltrated by the *lateral migratory stream*)

Endopiriform nucleus

Lateral ventricle

Layer VII (subplate)
Cortical plate
Layer I

Primary olfactory cortex (piriform)

Lateral nucleus?

Ammon's horn

Crus I ansiform lobule (HVIIA)

Future orbital gyri

Lateral fissure

Basal nucleus

HIPPOCAMPUS

Simplex lobule (HVI)

CEREBELLUM (HEMISPHERE)

AMYGDALA Cortical nuclei

Basal accessory nucleus

Subiculum

Entorhinal cortex

Primary fissure

Crus II ansiform lobule (HVIIA)

Lateral olfactory tract

Anterior lobe (HI-V)

Nucleus of the lateral lemniscus (dorsal)

CEREBELLUM (DEEP NUCLEI)

Middle cerebellar peduncle

Lateral lemniscus

Parabrachial nucleus

FRONTAL LOBE

Subiculum

Medial nucleus

Entorhinal cortex

Transpontine corticofugal tract

Principal sensory nucleus (V)

Superior cerebellar peduncle

Superior olivary complex (periolivary area)

CEREBELLUM (VERMIS)

Medial lemniscus

PONS

Raphe nuclear complex

Tuber (VIIb)?

Olfactory peduncle

Fourth ventricle

Pyramis (VII)

Uvula (IX)

Nodulus (X)

Subcallosal area

Pontocerebellar fibers

Pontocerebellar fibers (decussation)

Reticular formation

Pontine G/EP

Germinal trigone

Cerebellar NEP

Pontine gray

Interhemispheric fissure

Trigeminal motor nucleus (V)

Rostral migratory stream

Anterior extramural migratory stream
(contains pontine gray and reticular tegmental neurons)

External germinal layer

Vestibular nuclear complex?

Orbitofrontal NEP and SVZ

Perifascicular GEP
(may get contributions from the *rostral migratory stream* as it enlarges over anterior part of the lateral olfactory tract)

Inferior cerebellar peduncle

Dentate nucleus (*unlaminated*)

Middle cerebellar peduncle

Corticomedial complex

Amygdalo-hippocampal area

AMYGDALA

Hippocampal NEP

Orbitofrontal stratified transitional field

Basolateral complex

Ammonic migration

Central nucleus

STF1 t1
STF1 t2
STF2
STF3a
STF3b
STF3c
STF4 t2
STF5
STF6

Lateral migratory stream
(invades primary olfactory cortex)

Amygdaloid neuroepithelium

Lateral migratory stream
(invades amygdala)

Temporal (granular) stratified transitional field (STF)

Lateral migratory stream
(percolates through uncinate fasciculus)

Honeycomb matrix

Subpial granular layer
(GEP)

Temporal NEP and SVZ

**PLATE 55A
CR 170 mm, GW 20
Y391-62
Horizontal
Section 596**

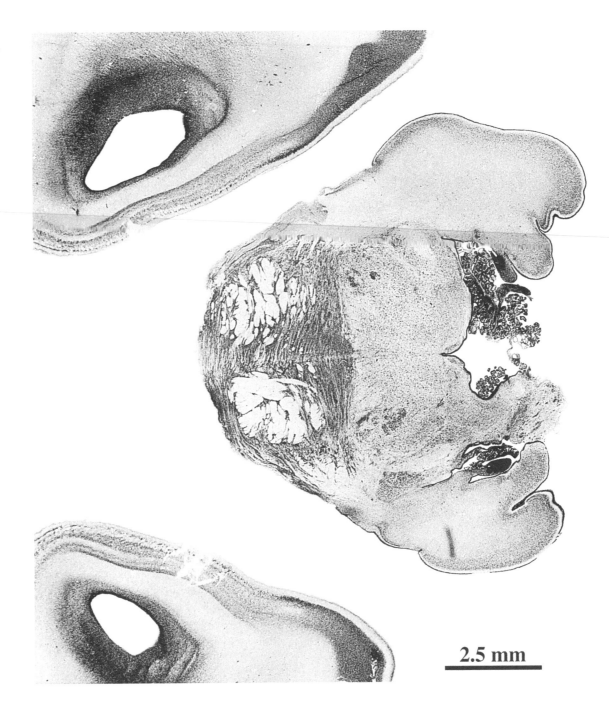

2.5 mm

**See the entire section
in Plates 48A and B.**

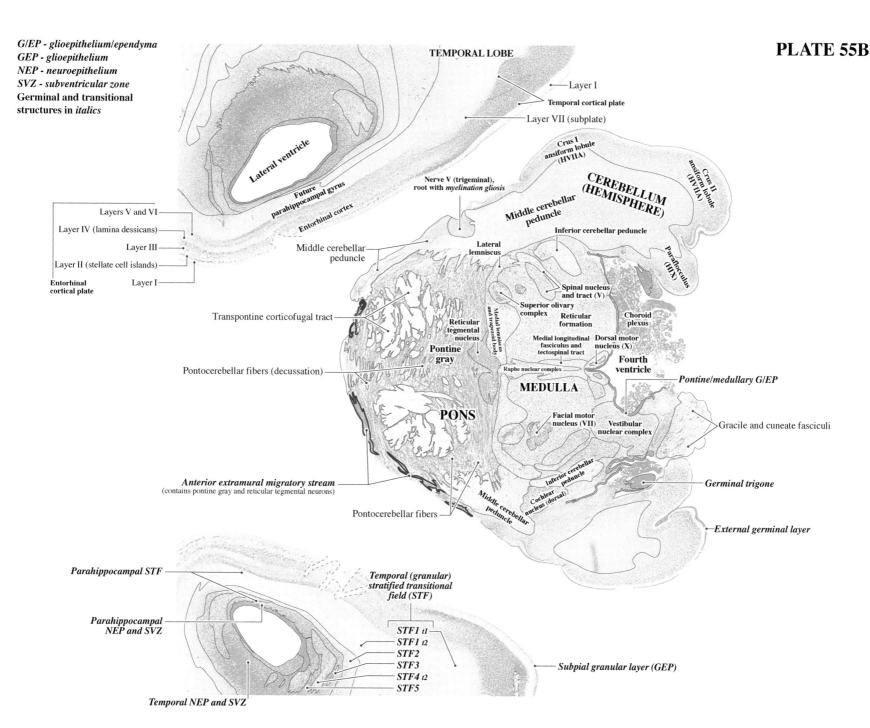

G/EP - glioepithelium/ependyma
GEP - glioepithelium
NEP - neuroepithelium
SVZ - subventricular zone
Germinal and transitional
structures in *italics*

TEMPORAL LOBE

Layer I
Temporal cortical plate
Layer VII (subplate)

Lateral ventricle

Crus I
ansiform lobule
(HVIIA)

Crus II
ansiform lobule
(HVIIA)

Future
parahippocampal gyrus

Nerve V (trigeminal),
root with *myelination gliosis*

CEREBELLUM
(HEMISPHERE)

Middle cerebellar
peduncle

Entorhinal cortex

Layers V and VI
Layer IV (lamina dessicans)
Layer III
Layer II (stellate cell islands)
Layer I
Entorhinal
cortical plate

Inferior cerebellar peduncle

Paraflocculus
(HX)

Lateral
lemniscus

Middle cerebellar
peduncle

Spinal nucleus
and tract (V)

Reticular
tegmental
nucleus

Superior olivary
complex

Reticular
formation

Choroid
plexus

Medial lemniscus
and trapezoid body

Transpontine corticofugal tract

Medial longitudinal
fasciculus and
tectospinal tract

Dorsal motor
nucleus (X)

Pontine
gray

Fourth
ventricle

Pontocerebellar fibers (decussation)

Raphe nuclear complex

MEDULLA

Pontine/medullary G/EP

PONS

Facial motor
nucleus (VII)

Vestibular
nuclear complex

Gracile and cuneate fasciculi

Anterior extramural migratory stream
(contains pontine gray and reticular tegmental neurons)

Inferior cerebellar
peduncle

Germinal trigone

Cochlear
nucleus (dorsal)

Pontocerebellar fibers

Middle cerebellar
peduncle

External germinal layer

Parahippocampal STF

*Temporal (granular)
stratified transitional
field (STF)*

*Parahippocampal
NEP and SVZ*

STF1 t1
STF1 t2
STF2
STF3
STF4 t2
STF5

Subpial granular layer (GEP)

Temporal NEP and SVZ

PLATE 56A
CR 170 mm, GW 20
Y391-62
Horizontal
Section 616

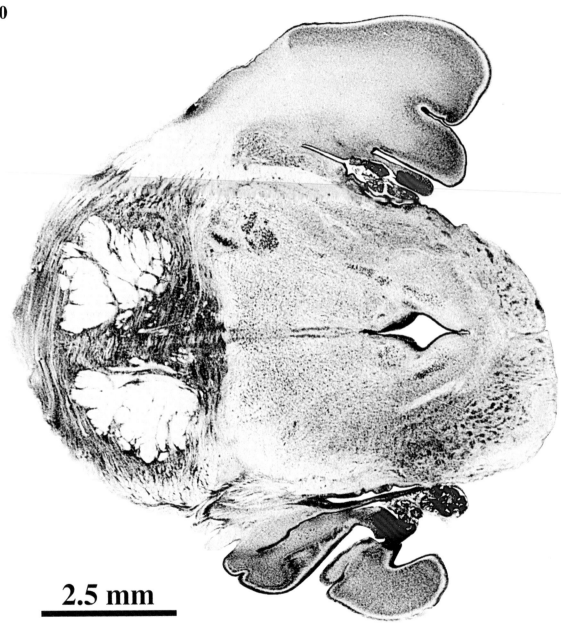

2.5 mm

G/EP - glioepithelium/ependyma
GEP - glioepithelium
Germinal and transitional
structures in *italics*

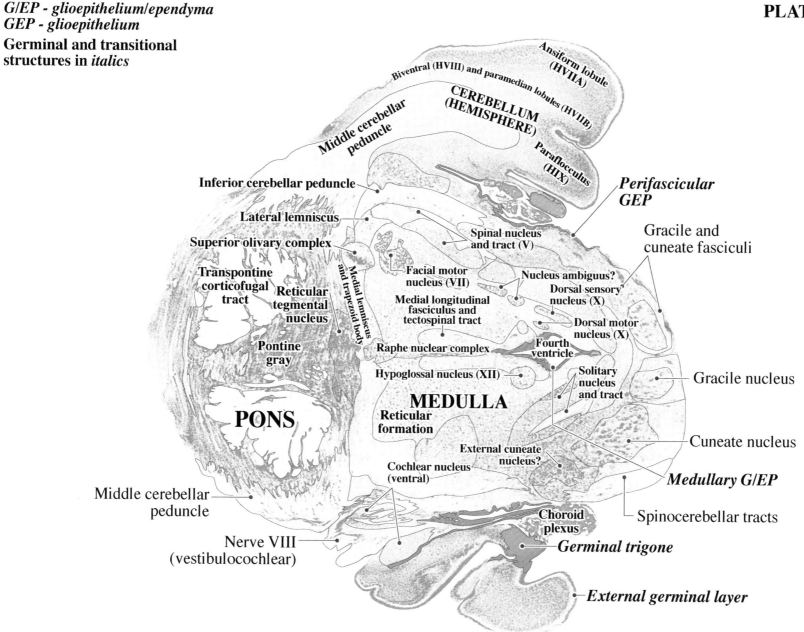

Ansiform lobule (HVIIA)

Biventral (HVIII) and paramedian lobules (HVIIB)

CEREBELLUM (HEMISPHERE)

Middle cerebellar peduncle

Paraflocculus (HIX)

Inferior cerebellar peduncle

Perifascicular GEP

Lateral lemniscus

Spinal nucleus and tract (V)

Gracile and cuneate fasciculi

Superior olivary complex

Transpontine corticofugal tract

Facial motor nucleus (VII)

Nucleus ambiguus?

Dorsal sensory nucleus (X)

Medial lemniscus and trapezoid body

Reticular tegmental nucleus

Medial longitudinal fasciculus and tectospinal tract

Dorsal motor nucleus (X)

Pontine gray

Raphe nuclear complex

Fourth ventricle

Solitary nucleus and tract

Gracile nucleus

Hypoglossal nucleus (XII)

MEDULLA

PONS

Reticular formation

Cuneate nucleus

External cuneate nucleus?

Medullary G/EP

Middle cerebellar peduncle

Cochlear nucleus (ventral)

Spinocerebellar tracts

Nerve VIII (vestibulocochlear)

Choroid plexus

Germinal trigone

External germinal layer

112

PLATE 57A
CR 170 mm, GW 20
Y391-62
Horizontal
Section 636

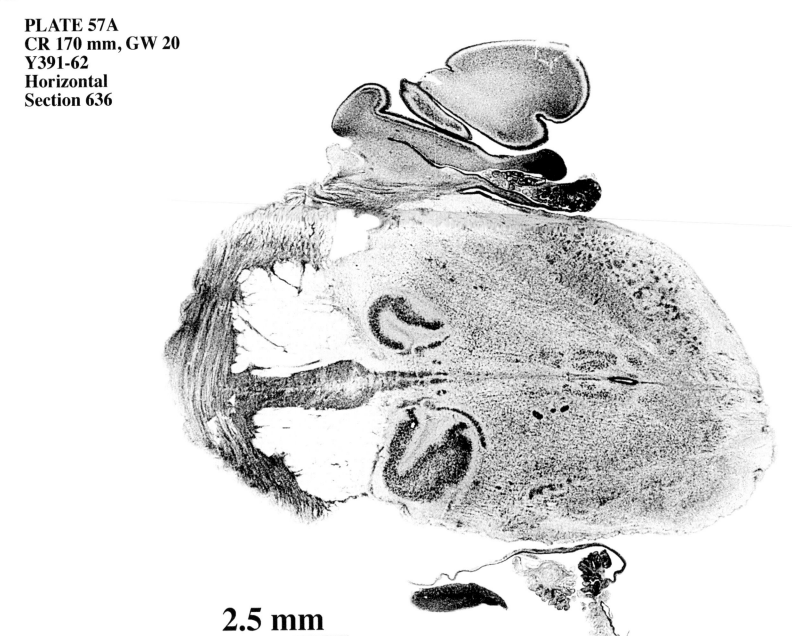

2.5 mm

G/EP - glioepithelium/ependyma
GEP - glioepithelium
Germinal and transitional
structures in *italics*

CEREBELLUM
(HEMISPHERE)

Biventral (HVIII) and
paramedian lobules (HVIIB)

External germinal layer

Flocculus (HX)

Cochlear
nucleus
(ventral)

Nerve VIII (vestibulocochlear)

Germinal trigone

Perifascicular GEP

Middle cerebellar
peduncle

Inferior cerebellar peduncle

Spinal nucleus
and tract (V)

Gracile and cuneate fasciculi

Cuneate nucleus

Solitary nucleus and tract

Nucleus ambiguus?

Inferior olive
(principal
nucleus)

Pontine
gray

Pyramid
(corticospinal tract)

Medial longitudinal
fasciculus and
tectospinal tract

Dorsal motor nucleus (X)

Dorsal sensory nucleus (X)

PONS

Medial lemniscus

Raphe nuclear complex

Fourth ventricle/central canal

Nucleus of Roller

Gracile nucleus

Medial
accessory
olive

Hypoglossal nucleus (XII)

MEDULLA

Spinal/
medullary
G/EP

Reticular
formation

Arcuate nucleus
(medulla)

Spinocerebellar tracts

Raphe migration

Choroid
plexus

Lateral reticular nucleus

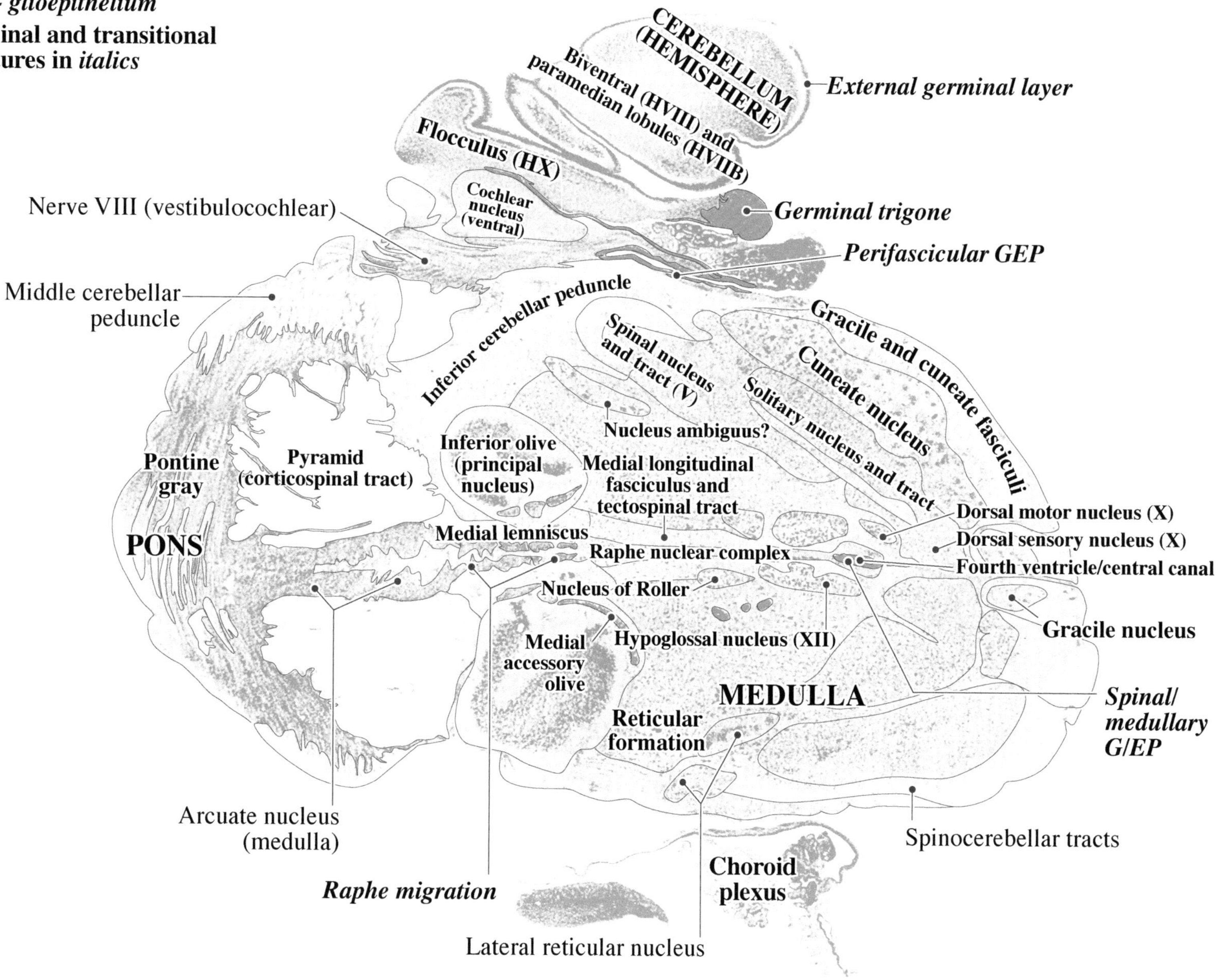

PLATE 58A
CR 170 mm, GW 20
Y391-62
Horizontal
Section 656

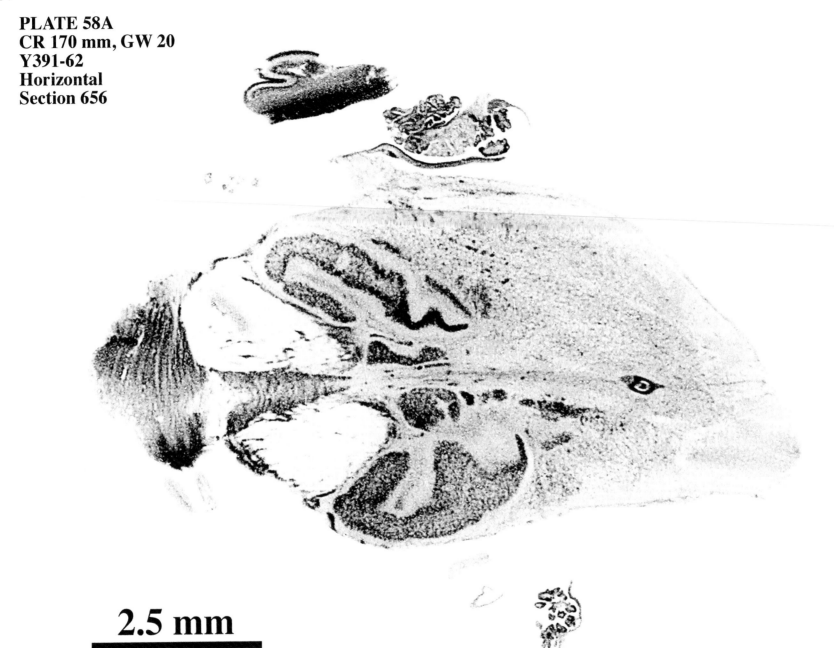

2.5 mm

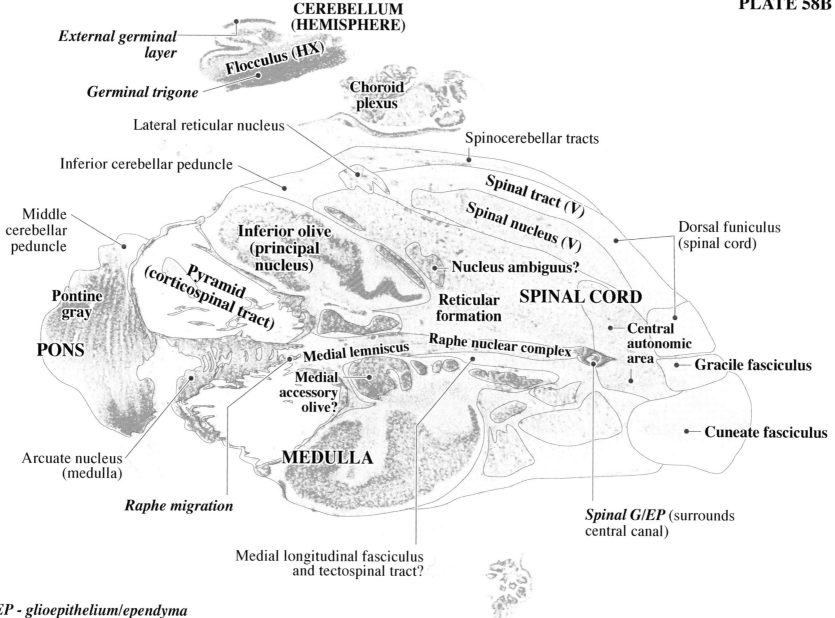

**CEREBELLUM
(HEMISPHERE)**

*External germinal
layer*

Flocculus (HX)

Germinal trigone

**Choroid
plexus**

Lateral reticular nucleus

Spinocerebellar tracts

Inferior cerebellar peduncle

Spinal tract (V)

Spinal nucleus (V)

Dorsal funiculus
(spinal cord)

Middle
cerebellar
peduncle

**Inferior olive
(principal
nucleus)**

Nucleus ambiguus?

**Pyramid
(corticospinal
tract)**

SPINAL CORD

**Pontine
gray**

**Reticular
formation**

**Central
autonomic
area**

PONS

Medial lemniscus

Raphe nuclear complex

Gracile fasciculus

**Medial
accessory
olive?**

Cuneate fasciculus

MEDULLA

Arcuate nucleus
(medulla)

Spinal G/EP (surrounds
central canal)

Raphe migration

Medial longitudinal fasciculus
and tectospinal tract?

G/EP - glioepithelium/ependyma

**Germinal and transitional
structures in *italics***

PLATE 59A
CR 170 mm, GW 20
Y391-62
Horizontal
Section 666

2.5 mm

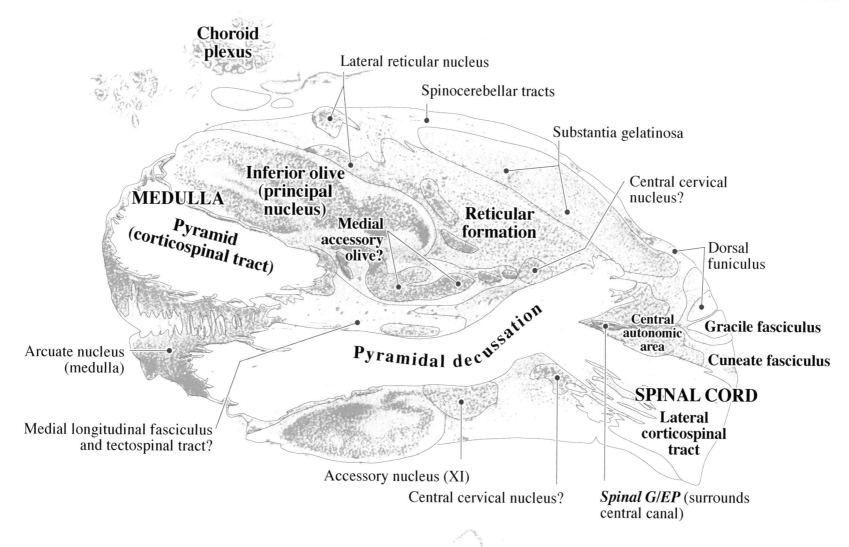

Choroid plexus

Lateral reticular nucleus

Spinocerebellar tracts

Substantia gelatinosa

Inferior olive (principal nucleus)

Central cervical nucleus?

MEDULLA

Pyramid (corticospinal tract)

Medial accessory olive?

Reticular formation

Dorsal funiculus

Pyramidal decussation

Arcuate nucleus (medulla)

Central autonomic area

Gracile fasciculus

Cuneate fasciculus

SPINAL CORD

Lateral corticospinal tract

Medial longitudinal fasciculus and tectospinal tract?

Accessory nucleus (XI)

Central cervical nucleus?

Spinal G/EP (surrounds central canal)

G/EP - glioepithelium/ependyma

Germinal and transitional structures in *italics*

T - #0294 - 160425 - C32 - 210/280/6 - PB - 9781032219424 - Gloss Lamination